Chemistry in the Community

ChemCom
Fourth Edition

CHEMISTRY
in the Community

CHEMCOM
Fourth Edition

A Project of the American Chemical Society

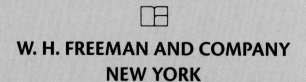

W. H. FREEMAN AND COMPANY
NEW YORK

ChemCom

Fourth Edition Credits

American Chemical Society

Chief Editor: Henry Heikkinen

Revision Team: Laurie Langdon, Robert Milne, Angela Powers

Revision Assistants: Cassie McClure, Seth Willis

Teacher Edition: Joseph Zisk, Lear Willis

Ancillary Materials: Regis Goode, Mike Clemente

Fourth Edition Editorial Advisory Board: Conrad L. Stanitski (Chair), Boris Berenfeld, Jack Collette, Robert Dayton, Ruth Leonard, Nina I. McClelland, George Milller, Adele Mouakad, Carlo Parravano, Kirk Soulé, Maria Walsh, Sylvia A. Ware (*ex officio*), Henry Heikkinen (*ex officio*)

ACS: Sylvia Ware, Janet Boese, Michael Tinnesand, Guy Belleman, Patti Galvan, Helen Herlocker

ACS Safety Committee: Henry Clayton Ramsey (Chair), Wayne Wolsey, Kevin Joseph Edgar, Herb Bryce.

W. H. Freeman

Publisher: Michelle Russel Julet

Project Editor: Rebecca Strehlow

Text Designer: Diana Blume

Cover Designers: Patricia McDermond, Diana Blume

Illustration Coordinator: Bill Page

Illustrations: Hudson River Studio

Photo Researcher: Elyse Reider

Production Coordinator: Susan Wein

Supplements and Multimedia Editor: Charlie Van Wagner

Composition: Black Dot Group

Layout: Proof Positive/Farrowlyne Associates

Manufacturing: R.R. Donnelly & Sons

This material is based upon work supported by the National Science Foundation under Grant No. SED-88115424 and Grant No. MDR-89470104. Any opinions, findings, and conclusions or recommendations expressed in this publication are those of the authors and do not necessarily reflect the views of the National Science Foundation. Any mention of trade names does not imply endorsement by the National Science Foundation.

NOTE TO STUDENTS

This is likely to be your first chemistry course. But without realizing it, you have been associated with chemistry for quite some time. Chemical reactions are all around you in your daily living, involved in foods, fuels, fabrics, and medicines. Chemistry is the study of substances in our world and is responsible for the materials that make up where you live, how you are transported, what you eat and wear, even you yourself.

As society depends more on technology, that is, the application of science, the decisions that individuals, communities, and nations make will rely more heavily on citizens—not just those who are scientists—to understand the scientific phenomena and principles required by such decisions. As a future voter, you can bring your chemical knowledge to decisions that require it. This textbook was developed to present ways for you to apply chemical concepts in order to help you understand the chemistry behind some important socio-technological issues.

Each of the units introduces issues that concern your life, your community, and the relation of chemistry to them. You will be involved in laboratory activities and other problem-solving exercises that ask you to apply your chemical knowledge to a particular problem. You will seek solutions and evaluate the consequences of those you propose, because chemistry applied to communities is not without consequences.

We begin with a newspaper article about a water-related emergency in Riverwood, an issue that is the theme for the entire first unit. Can chemistry help solve the problem? Welcome to *ChemCom* and to Riverwood!

SAFETY AND
LABORATORY ACTIVITIES

IMPORTANT NOTICE

ChemCom is intended for use by high school students in the classroom laboratory under the direct supervision of a qualified chemistry teacher. The experiments described in this book involve substances that may be harmful if they are misused or if the procedures described are not followed. Read cautions carefully, and follow all directions. Do not use or combine any substances or materials not specifically called for in carrying out experiments. Other substances are mentioned for educational purposes only and should not be used by students unless the instructions specifically so indicate.

The materials, safety information, and procedures contained in this book are believed to be reliable. This information and these procedures should serve only as a starting point for good laboratory practices, and they do not purport to specify minimal legal standards or to represent the policy of the American Chemical Society. No warranty, guarantee, or representation is made by the American Chemical Society as to the accuracy or specificity of the information contained herein, and the American Chemical Society assumes no responsibility in connection therewith. The added safety information is intended to provide basic guidelines for safe practices. It cannot be assumed that all necessary warnings and precautionary measures are contained in the document and that other additional information and measures may not be required.

In *ChemCom* you will frequently perform laboratory activities. While no human activity is completely risk free, if you use common sense, as well as chemical sense, and follow the rules of laboratory safety, you should encounter no problems. Chemical sense is an extension of common sense. Sensible laboratory conduct won't happen by memorizing a list of rules, any more than a perfect score on a written driver's test ensures an excellent driving record. The true "driver's test" of chemical sense is your actual conduct in the laboratory.

The following safety pointers apply to all laboratory activities. For your personal safety and that of your classmates, make following these guidelines second nature in the laboratory. Your teacher will point out any special safety guidelines that apply to each activity. Three safety icons appear in your textbook. They appear at the beginning of the laboratory exercise, but apply to the entire lab activity. When you see the goggle icon you should put on your safety goggles and continue to wear them on your eyes until you are completely finished with the lab. A lab apron icon appears when there is a spill hazard, and you should wear your lab coat or apron. The caution icon means there are substances or procedures that require special care. See your teacher for specific information on these cautions.

If you understand the reasons behind them, these safety rules will be easy to remember and to follow. So, for each listed safety guideline:

- Identify a similar rule or precaution that applies in everyday life— for example in cooking, repairing or driving a car, or playing a sport.

- Briefly describe possible harmful results if the rule is not followed.

Rules of Laboratory Conduct

1. Perform laboratory work only when your teacher is present. Unauthorized or unsupervised laboratory experimenting is not allowed.

2. Your concern for safety should begin even before the first activity. Before starting the laboratory activity, always read and think about each laboratory assignment.

3. Know the location and use of all safety equipment in your laboratory. These should include the safety shower, eye wash, first-aid kit, fire extinguisher, fire blanket, exits, and evacuation routes.

4. Wear a laboratory coat or apron and impact/splash-proof goggles for all laboratory work. Wear shoes (rather than sandals or open-toed shoes), and tie back loose hair. Shorts or short skirts must not be worn.

5. Clear your benchtop of all unnecessary material such as books and clothing before starting your work.

6. Check chemical labels twice to make sure you have the correct substance and the correct concentration of a solution. Some chemical formulas and names may differ by only a letter or a number.

7. You may be asked to transfer some laboratory chemicals from a common bottle or jar to your own container. Do not return any excess material to its original container unless authorized by your teacher, as you may contaminate the common bottle.

8. Avoid unnecessary movement and talk in the laboratory.

9. Never taste laboratory materials. Do not bring gum, food, or drinks into the laboratory. Do not put fingers, pens, or pencils in your mouth while in the laboratory.

10. If you are instructed to smell something, do so by fanning some of the vapor toward your nose. Do not place your nose near the opening of the container. Your teacher will show the correct technique.

11. Never look directly down into a test tube; do view the contents from the side. Never point the open end of a test tube toward yourself or your neighbor. Never heat a test tube directly in a Bunsen burner flame.

12. Any laboratory accident, however small, should be reported immediately to your teacher.

13. In case of a chemical spill on your skin or clothing rinse the affected area with plenty of water. If the eyes are affected, rinsing with water must begin immediately and continue for at least 10 to 15 minutes. Professional assistance must be obtained.

14. Minor skin burns should be placed under cold, running water.

15. When discarding or disposing of used materials, carefully follow the instructions provided.

16. Return equipment, chemicals, aprons, and protective goggles to their designated locations.

17. Before leaving the laboratory, make sure that gas lines and water faucets are shut off.

18. Wash your hands before leaving the laboratory.

19. If you are unclear or confused about the proper safety procedures, ask your teacher for clarification. If in doubt, ask!

Safety and Laboratory Activities

CONTENTS

UNIT 3 PETROLEUM: BREAKING AND MAKING BONDS   172

UNIT 4 AIR: CHEMISTRY AND THE ATMOSPHERE

UNIT 4 ADDRESSES the chemistry of gaseous substances found in Earth's atmosphere. Chemistry content includes exploration and application of the gas laws and an introduction to the kinetic molecular theory and the concept of electromagnetic radiation. Ideas about acids, bases, and buffers are developed in the context of acid rain. The chemistry of common air pollutants, including particulates, smog, and ozone, is introduced and related to important community and global issues.

The new edition of this unit includes added material on solar radiation and the electromagnetic spectrum and expanded attention to acids, bases, and buffers. The new edition also includes some reordering of unit topics for added clarity and continuity.

UNIT 5 INDUSTRY: APPLYING CHEMICAL REACTIONS

UNIT 5 FOCUSES on electrochemistry and the chemistry of nitrogen in the context of industrial chemistry's role in modern society. The nitrogen cycle and nitrogen fixation are explored, and oxidation states and electronegativity are used to aid understanding of a range of important chemical reactions.

The new edition of this unit includes increased emphasis on kinetics, equilibria, and catalysis and also revisits the idea of benefit-burden analysis. Additionally, implications of green chemistry help clarify contemporary decision-making within chemical industries.

UNIT 6 ATOMS: NUCLEAR INTERACTIONS

UNIT 6 FOCUSES on the energy and interactions associated with atomic nuclei. The chemistry of isotopes is introduced in relation to radioactivity and radioactive decay. The effects, uses, and challenges of ionizing radiation and key properties of different kinds of radiation are explored. Finally, the role of nuclear energy and the challenges of nuclear waste disposal are considered in societal and scientific contexts.

The new edition of this unit includes up-to-date information on synthetically produced transfermium elements and expanded coverage of nuclear applications in medical diagnosis.

UNIT 7 FOOD: MATTER AND ENERGY FOR LIFE

THE FOCUS OF UNIT 7 is on structures, properties, and reactions of major food components—specifically, fats, carbohydrates, and proteins. The structures and functions of vitamins, minerals, and food additives are also explored. The notion of limiting reactants is used to help clarify the roles of essential molecular and mineral nutrients.

The new edition of this unit features added material on enzyme chemistry and a more detailed analysis of cellular energy flow. Chemical bonding is also featured, reinforcing ideas on bond breaking and forming as well as the energy involved in those processes.

Welcome to
CHEMISTRY IN THE COMMUNITY

Since 1988, when the first edition of *Chemistry in the Community* (*ChemCom*) was published, *ChemCom* has been successfully used by more than one million students in many different types of high schools. Developed by the American Chemical Society (ACS) with financial support from the National Science Foundation and several ACS sources, the writing of this edition was assisted by an Editorial Advisory Board composed of high school chemistry teachers, university chemistry professors, and chemists from industry and federal agencies.

In brief, the goals of *ChemCom* are to help you

♦ recognize and understand the importance of chemistry in your life;

♦ develop problem-solving techniques and critical-thinking skills that will enable you to apply chemical principles in making informed decisions about scientific and technological issues; and

♦ acquire an awareness of the potential as well as the limitations of science and technology.

As in the previous editions of *ChemCom*, chemical principles are presented on a "need-to-know" basis. Each of the seven units begins with a significant socio-technological issue that provides the framework around which the appropriate chemistry is introduced and developed. Woven into this tapestry of chemistry are chemical principles relevant to that particular socio-technological issue. Each unit deals in some way with a community and its chemistry, that community possibly taking the form of school, town, region, country—perhaps even Planet Earth. And as before, each unit ends with a consolidating activity. Laboratory activities continue to be integral parts of each unit, as do decision-making opportunities and skill-building exercises.

The following pages will introduce you to the various features of this textbook. Great care has been taken to present the concepts of chemistry in a way that facilitates your learning. This introduction is intended to support one of your roles in the learning process: knowing how to use your textbook effectively.

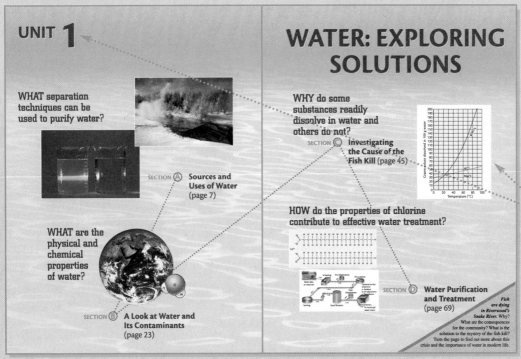

UNIT 1

WATER: EXPLORING SOLUTIONS

WHAT separation techniques can be used to purify water?

SECTION Ⓐ Sources and Uses of Water (page 7)

WHAT are the physical and chemical properties of water?

SECTION Ⓑ A Look at Water and Its Contaminants (page 23)

WHY do some substances readily dissolve in water and others do not?

SECTION Ⓒ Investigating the Cause of the Fish Kill (page 45)

HOW do the properties of chlorine contribute to effective water treatment?

SECTION Ⓓ Water Purification and Treatment (page 69)

Fish are dying in Riverwood's Snake River. Why? What are the consequences for the community? What is the solution to the mystery of the fish kill? Turn the page to find out more about this crisis and the importance of water in modern life.

Unit Openers provide a visual preview of the material to come. The images and motivating questions are designed to generate interest and stimulate thinking about the theme of the unit. How much do you already know? What are you about to discover?

The first four units are designed to be studied in sequence because each unit builds upon the previous ones, reinforcing the idea that the concepts of chemistry are interrelated and applicable to a variety of situations. The last three units can be studied in any sequence.

The seven units of *ChemCom* are:
Water: Exploring Solutions
Materials: Structure and Uses
Petroleum: Breaking and Making Bonds

Air: Chemistry and the Atmosphere
Industry: Applying Chemical Reactions
Atoms: Nuclear Interactions
Food: Matter and Energy for Life

You are introduced to each unit through a compelling socio-technological issue. Whether it is a fish kill of crisis proportion in the town of Riverwood or a challenge to design a prototypical Moon base, each scenario places the content in a relevant context. Over and over again, you will return to the issue as you gather chemical information, participate in hands-on learning, make decisions, and strengthen your chemistry-based problem-solving skills. At the conclusion of the unit, you will have an opportunity to put together all that you have learned to address the issue.

Each unit consists of several lettered sections followed by a concluding (unlettered) section. This last section is the culminating activity titled Putting It All Together. Sections are further divided into numbered topics and features. You can skim the section and get a sense of its organization by reading the subsection titles. If you organize the information you learn, your understanding of it will be greatly enhanced.

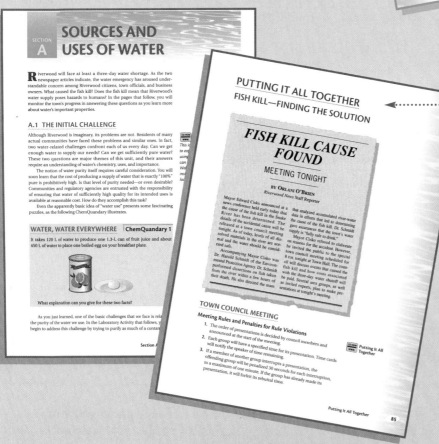

Learning the processes of science is as important as learning its results. Laboratory activities give you an opportunity to experiment in much the same way a chemist does. Through these motivating, hands-on activities, you will learn important chemical principles, practice laboratory skills, develop critical thinking, and use data and observations to draw conclusions.

The laboratory is a place of discovery and excitement—a place in which you will gain first-hand knowledge about the processes of science and chemical phenomena. To make your experience in the chemistry laboratory safe and rewarding, it is important that you follow certain basic safety precautions. The safety icons throughout the textbook remind you to exercise appropriate caution. Before you perform your first laboratory activity, carefully read the safety information on pages iv–vi.

Data collection is an important part of any laboratory procedure. For many of the laboratory activities, sample Data Tables have been provided. As you become more experienced in laboratory work, you will have less need for a sample table; devising an appropriate Data Table will become part of your laboratory skills.

Laboratory activities vary in length and complexity. However, each activity usually consists of an Introduction, a detailed or student-developed Procedure, and Questions. The activity may also contain Calculations, Data Analysis, and Post-Lab Activities.

All of the terms introduced in *ChemCom* are important to your understanding of chemistry. There are some terms, however, that deserve additional emphasis and therefore appear in boldface type. You need not memorize these terms; rather, you should be able to explain their meanings, use them correctly in scientific contexts, and describe how they are interrelated.

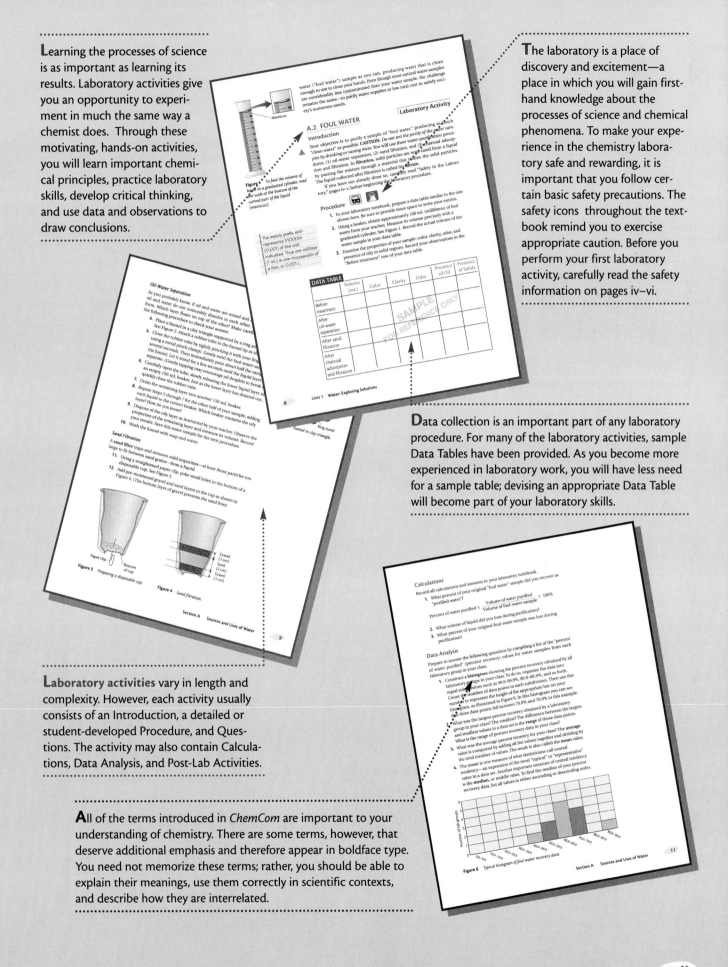

Many of the issues you will face as an informed citizen will require decisions, including science-based ones. The more practiced you are at decision making, the wiser your decisions are apt to be. Making Decisions provides opportunities for you to gain experience with real-life decision-making strategies—often in a cooperative-group setting.

In each Making Decisions, you will be presented with aspects of societal/technological problems, asked to collect and/or analyze data for underlying patterns, and challenged to develop and evaluate answers based on scientific evidence and potential consequences.

Understanding the processes involved in solving a problem can be as important as obtaining the answer. The Building Skills activities found throughout ChemCom can help you strengthen your problem-solving skills. These real-world chemistry situations and practice problems provide reinforcement of basic chemical concepts, skills, and calculations.

Chem Quandary

These activities allow you to use scientific concepts to help frame and answer interesting questions. Challenging and motivational, they encourage open-ended thought and may generate additional questions about chemistry and society.

Interesting, relevant, and sometimes amusing information can be found in margin notes distributed throughout the textbook. Often these notes serve as reminders of material previously presented or previews of topics to come.

The **Modeling Matter** activities make abstract chemical concepts easier to grasp by helping you visualize or "see" the fundamental structures and interactions of atoms. In doing these activities, you will be critiquing and creating visual respresentations of chemical activity, formulating and revising scientific explanations, developing manipulable models, and using logic and evidence to make informed decisions.

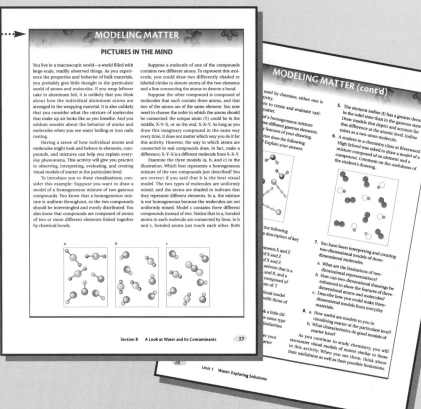

"**D**oing chemistry" is not limited to work performed on the surface of a laboratory benchtop. Chemistry is used by a wide variety of people engaged in a broad range of careers. You will meet some of these people and learn how their work is an application of the processes and principles of chemistry as you read **Chemistry At Work.** Perhaps you will be encouraged to find out on your own more about particular areas of study and careers that incorporate chemistry.

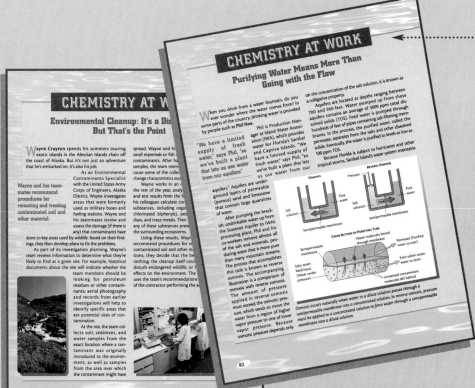

Welcome to Chemistry in the Community

Each lettered section in a unit concludes with a **Section Summary.** Here you will have an opportunity to review the important concepts you have learned and skills you have developed. The summary is organized around each section's learning goals: broad statements that identify what you should have learned and provide a context for the questions you are to answer.

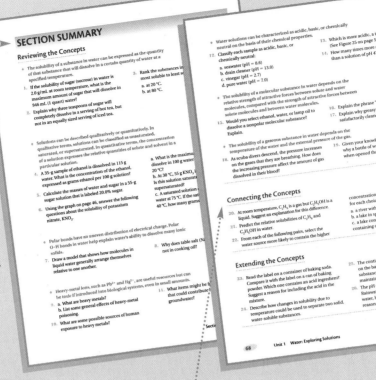

The Section Summary is divided into three categories of questions. **Reviewing the Concepts** allows for practice of skills and recall of fundamental concepts. In **Connecting the Concepts** you will identify and develop relationships among the various facts and ideas. Finally, **Extending the Concepts** challenges you to seek additional information and apply the chemistry you have learned to new situations.

You will learn a great deal about chemistry in each section of a unit. It is important that you sum up, or consolidate, the information you have learned so that it is meaningful and easily retained. The closing part of each unit, **Putting It All Together** gives you the opportunity to review, integrate, and apply what you have learned in the context of a chemistry-related real-world issue. You will be expected to develop and defend scientifically sound positions that acknowledge the roles of economics, politics, and personal/social values. Working individually or in small groups, you will weigh the risks and benefits of your decisions and then share, compare, or negotiate those decisions with others in your class.

CHEMCOM MEDIA: USING THE INTERNET AND OTHER TECHNOLOGY

In only a very short time, the Internet has become a leading source of information and technology. Five years ago, people knew the World Wide Web existed, but few actually reaped its benefits because of dated hardware, slow connections, and slow downloads. Now, with more innovation and better technology, just about everyone is "connected." As a result of recent breakthroughs, businesses, governments, and even schools are taking advantage of this vast information system. With a good search engine and a fast connection, the Internet can be a valuable and indispensable tool or aid in research or data collection.

Companies are now taking the current technology one step further by installing teaching programs on the Web. By setting up distance-learning programs complete with interactivity, universities and businesses are helping students like you master subjects over the Internet.

W. H. Freeman and Company has joined this technological revolution by creating study aides that are readily accessible and easily navigable. For the first time, *Chemistry in the Community* (*ChemCom*), now in its fourth edition, comes with an electronic component developed by W. H. Freeman. Currently available on the Web are the first three units from the *ChemCom CD-ROM,* which contains animations, videos, interactive exercises, study tools, and assessment opportunities—all designed to enable students to delve deeper into difficult concepts and to test their knowledge.

ChemCom CD-ROM is designed to enhance the textbook, not to replace it. The software is exciting, entertaining, and simple to use. You will have fun while learning about chemistry.

In order to gain access to *ChemCom CD-ROM* on the Web, you must dial into the Internet on the server you currently use and key "www.whfreeman.com/chemcom" into the address box at the top of your screen. This will bring you directly to the *ChemCom* site. You can navigate by clicking on the section you are interested in and walk through all of the features that apply to that section by clicking "next" at the bottom right of your screen. Follow the directions on the screen when you are playing animations and videos or answering questions through the interactive exercises. After you are finished browsing, test your knowledge on the Question & Answer page located at the bottom left of your screen.

The following pages are illustrated with screen shots from *ChemCom CD-ROM.* Note the exciting features that are highlighted and the instructions provided. They will help you successfully navigate and use the site.

If you have any problems using *ChemCom CD-ROM* on the Web contact our tech support at:

Techsupport@bfwpub.com
Or call 1-800-936-6899

 Overview in the margin of the book
This icon indicates an opportunity to explore further resources by using the ChemCom CD-Rom. You can also access the resources via the Web. See your teacher for further instructions. The appropriate location is printed near each icon.

UNIT OPENER

Tells you where you are in your navigation

Tells you the available media

Navigate by section

Transports you to study aids such as calculators and an interactive Periodic Table

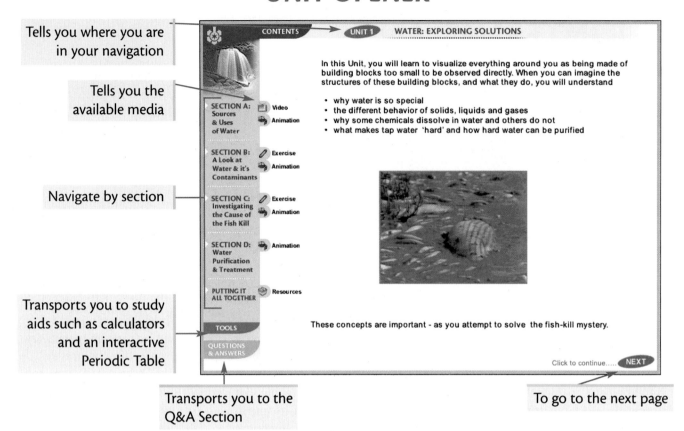

CONTENTS → UNIT 1 WATER: EXPLORING SOLUTIONS

In this Unit, you will learn to visualize everything around you as being made of building blocks too small to be observed directly. When you can imagine the structures of these building blocks, and what they do, you will understand

- why water is so special
- the different behavior of solids, liquids and gases
- why some chemicals dissolve in water and others do not
- what makes tap water 'hard' and how hard water can be purified

These concepts are important - as you attempt to solve the fish-kill mystery.

Click to continue..... NEXT

SECTION A: Sources & Uses of Water — Video, Animation

SECTION B: A Look at Water & it's Contaminants — Exercise, Animation

SECTION C: Investigating the Cause of the Fish Kill — Exercise, Animation

SECTION D: Water Purification & Treatment — Animation

PUTTING IT ALL TOGETHER — Resources

TOOLS

QUESTIONS & ANSWERS

Transports you to the Q&A Section

To go to the next page

SECTION OPENER

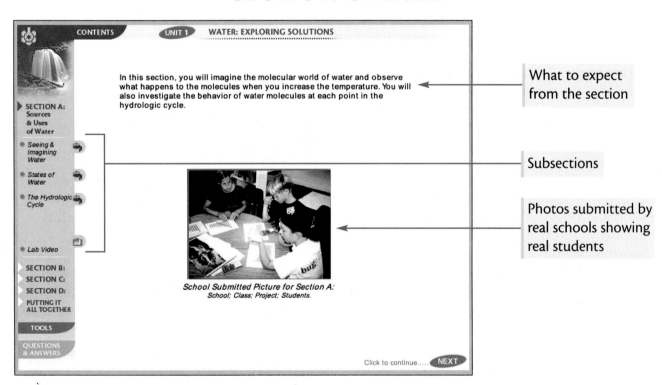

CONTENTS UNIT 1 WATER: EXPLORING SOLUTIONS

In this section, you will imagine the molecular world of water and observe what happens to the molecules when you increase the temperature. You will also investigate the behavior of water molecules at each point in the hydrologic cycle.

What to expect from the section

SECTION A: Sources & Uses of Water
- Seeing & Imagining Water
- States of Water
- The Hydrologic Cycle
- Lab Video

SECTION B:
SECTION C:
SECTION D:
PUTTING IT ALL TOGETHER

TOOLS

QUESTIONS & ANSWERS

Subsections

Photos submitted by real schools showing real students

School Submitted Picture for Section A:
School; Class; Project; Students.

Click to continue..... NEXT

ANIMATIONS

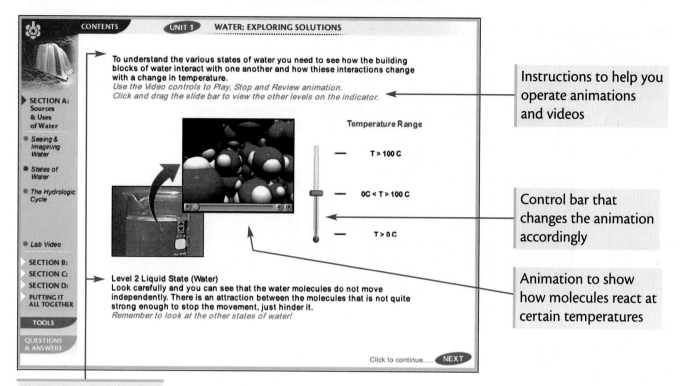

Instructions to help you operate animations and videos

Control bar that changes the animation accordingly

Animation to show how molecules react at certain temperatures

Text to explain what is on the screen

DIAGRAMS

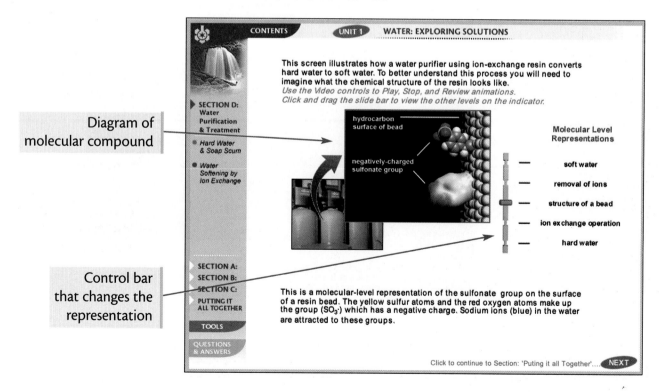

Diagram of molecular compound

Control bar that changes the representation

INTERACTIVE EXERCISES

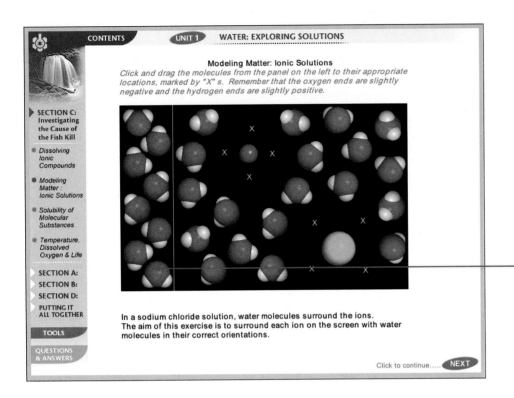

Before the exercise

Drag these molecules to the appropriate place

After the exercise

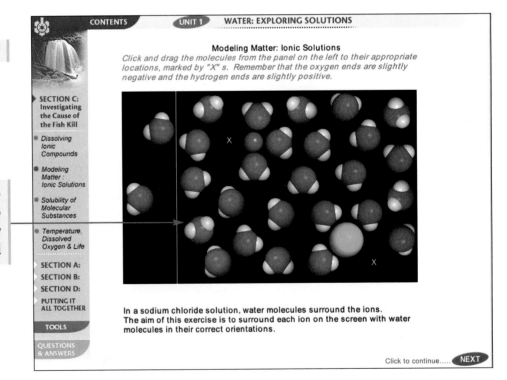

Drag and Drop Exercises that help you understand how molecules are formed

INTERACTIVE EXERCISES

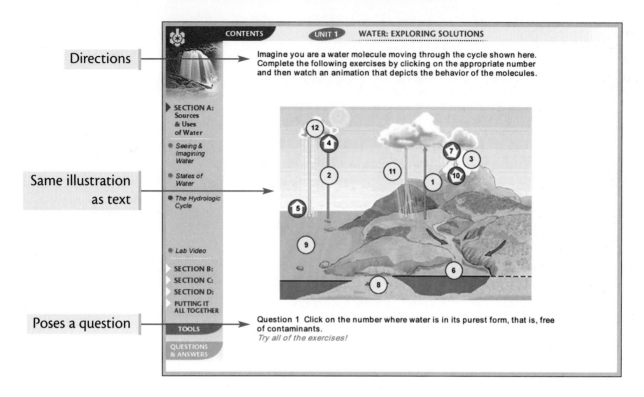

Directions

Same illustration as text

Poses a question

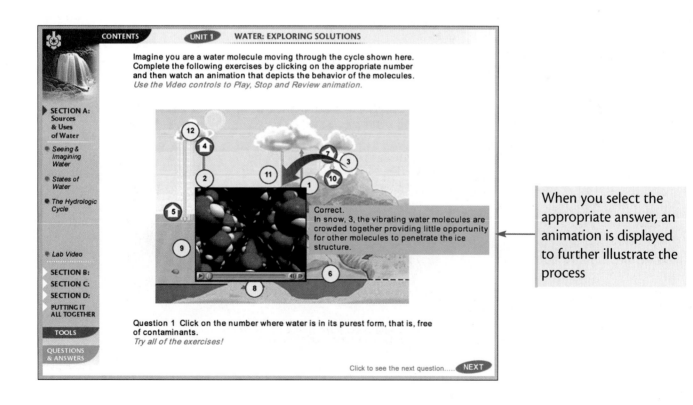

When you select the appropriate answer, an animation is displayed to further illustrate the process

UNIT 1

WHAT separation
techniques can be
used to purify water?

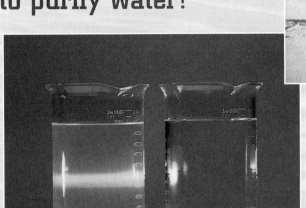

WHAT are the
physical and
chemical
properties
of water?

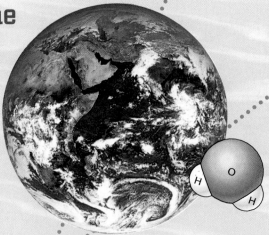

WATER: EXPLORING SOLUTIONS

WHY do some
substances readily
dissolve in water and
others do not?

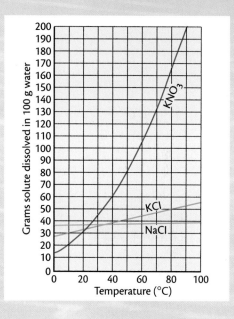

HOW do the properties of chlorine
contribute to effective water treatment?

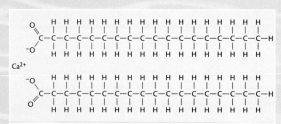

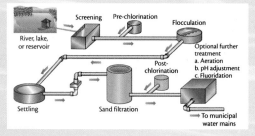

*Fish
are dying
in Riverwood's
Snake River.* Why?
What are the consequences
for the community? What is the
solution to the mystery of the fish kill?
Turn the page to find out more about this
crisis and the importance of water in modern life.

FISH KILL CAUSES RIVERWOOD WATER EMERGENCY

SEVERE WATER RATIONING IN EFFECT

BY LORI KATZ

Riverwood News Staff Reporter

Citing possible health hazards, Mayor Edward Cisko announced today that Riverwood will stop withdrawing water from the Snake River and will temporarily shut down the city's water-treatment plant. Commencing at 6 P.M., river water will not be pumped to the plant for at least three days. And, if the cause of the fish kill has not been determined and corrected by that time, the shutdown will continue indefinitely.

During the pumping-station shutdown, water engineers and chemists from the County Sanitation Commission and Environmental Protection Agency (EPA) will investigate the cause of the major fish kill discovered yesterday. The fish kill extended from the base of Snake River Dam, located upstream from Riverwood, to the town's water-pumping station.

The initial alarm was sounded when Jane Abelson, 15, and Chad Wong, 16—both students at Riverwood High School—found many dead fish floating in a favorite fishing spot. "We thought maybe someone had poured poison into the reservoir," explained Wong.

Mary Steiner, Riverwood High School biology teacher, accompanied the students back to the river. "We hiked downstream for almost a mile. Dead fish of all kinds were washed up on the banks and caught in the rocks," Abelson reported.

Ms. Steiner contacted County Sanitation Commission officers, who immediately collected Snake River water samples for analysis. Chief Engineer Hal Cooper reported at last night's emergency meeting that the water samples appeared clear and odorless. However, he indicated some concern. "We can't say for sure that our water supply is safe until the reason for the fish kill is known. It's far better that we take no chances until then," Cooper advised.

Mayor Cisko canceled the community's "Fall Fish-In," scheduled to start Friday. No plans for rescheduling Riverwood's annual fishing tournament were announced. "Consensus at last night's emergency town council meeting was to start an investigation of the situation immediately," he said.

After five hours of often heated debate yesterday, the town council finally reached agreement to stop drawing water from the Snake River. Council member Henry McLatchen (also a Chamber of Commerce member) said the decision was highly emotional and unnecessary. He cited financial losses that motels and restaurants will suffer because of the

see Fish Kill, page 5

Fish Kill, from page 4

Fish-In cancellation, as well as potential loss of future tourism dollars due to adverse publicity. However, McLatchen and other council members sharing that view were outvoted by the majority, who expressed concern that the fish kill—the only one to occur in the Riverwood region in the past 30 years—may indicate a public health emergency.

Mayor Cisko assured residents that essential municipal services will not be affected by the crisis. Specifically, he promised to maintain fire department access to adequate supplies of water to meet fire-fighting needs.

Arrangements have been made to truck emergency drinking water from Mapleton. First water shipments are due to arrive in Riverwood by mid-morning tomorrow. Distribution points are listed in Part 2 of today's *Riverwood News*, along with guidelines on saving and using water during this emergency.

All Riverwood schools will be closed Monday through Wednesday. No other closings or cancellations have yet been announced. Local TV and radio will report new schedule changes as they are released.

A public meeting tonight at 8 P.M. at Town Hall features Dr. Margaret Brooke, a water expert at State University. She will answer questions concerning water safety and use. Brooke was invited by the County Sanitation Commission to help clarify the situation for concerned citizens.

Asked how long the water emergency would last, Dr. Brooke refused any speculation, saying that she first needed to talk to chemists conducting the investigation. EPA scientists, in addition to collecting and analyzing water samples, will examine dead fish in an effort to determine what was responsible for the killing. She reported that trends or irregularities in water-quality data from Snake River monitoring during the past two years also will play a part in the investigation.

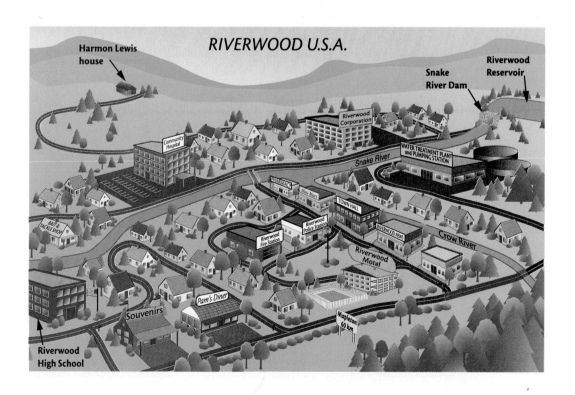

RIVERWOOD U.S.A.

TOWNSPEOPLE REACT TO FISH KILL AND ENSUING WATER CRISIS

BY JUAN HERNANDEZ
Riverwood News Staff Reporter

In a series of on-the-street interviews, Riverwood citizens expressed a variety of feelings earlier today about the crisis. "It doesn't bother me," said nine-year-old Jimmy Hendricks. "I'm just going to drink bottled water and pre-packaged fruit juice."

"I knew that eventually they'd pollute the river and kill the fish," complained Harmon Lewis, a lifelong resident of Fieldstone Acres, located east of Riverwood. Lewis, who traces his ancestry to original county settlers, still gets his water from a well and will be unaffected by the water crisis. He said that he plans to pump enough well water to supply the children's ward at Community Hospital if the emergency extends more than a few days.

Bob and Ruth Hardy, owners of Hardy's Ice Cream Parlor, expressed annoyance at the inconvenience but were reassured by council actions. They were eager to learn the reason for the fish kill and its possible effects on future water supplies.

The Hardy's daughter Toni, who loves to fish, was worried that late-season fishing would be ruined. Toni and her father won first prize in last year's angling competition.

Riverwood Motel owner Don Harris expressed concern for both the health of town residents and the loss of business due to the tournament cancellation. "I always earn reasonable income from this event and, without the revenue from the Fish-In, I may need a loan to pay bills in the spring."

The unexpected school vacation was "great," according to twelve-year-old David Price. Asked why he thought schools would be closed during the water shortage, Price said that all he could think of was that "the drinking fountains won't work."

Elmo Turner, whose residential landscaping has won Garden Club recognition for the past five years, felt reassured on one point. Because of the unusually wet summer, grass watering is unnecessary, and lawns are in no danger of drying as a result of current water rationing.

Dead fish were found washed up along the banks of the Snake River yesterday afternoon.

SOURCES AND USES OF WATER

Riverwood will face at least a three-day water shortage. As the two newspaper articles indicate, the water emergency has aroused understandable concern among Riverwood citizens, town officials, and business owners. What caused the fish kill? Does the fish kill mean that Riverwood's water supply poses hazards to humans? In the pages that follow, you will monitor the town's progress in answering these questions as you learn more about water's important properties.

A.1 THE INITIAL CHALLENGE

Although Riverwood is imaginary, its problems are not. Residents of many actual communities have faced these problems and similar ones. In fact, two water-related challenges confront each of us every day. Can we get enough water to supply our needs? Can we get sufficiently pure water? These two questions are major themes of this unit, and their answers require an understanding of water's chemistry, uses, and importance.

The notion of water purity itself requires careful consideration. You will soon learn that the cost of producing a supply of water that is exactly "100%" pure is prohibitively high. Is that level of purity needed—or even desirable? Communities and regulatory agencies are entrusted with the responsibility of ensuring that water of sufficiently high quality for its intended uses is available at reasonable cost. How do they accomplish this task?

Even the apparently basic idea of "water use" presents some fascinating puzzles, as the following ChemQuandary illustrates.

Overview

This icon indicates an opportunity to explore further resources by using the ChemCom CD-Rom. You can also access the resources via the Web. See your teacher for further instructions. The appropriate location is printed near each icon.

Sources & Uses of Water

WATER, WATER EVERYWHERE ⋮ ChemQuandary 1

It takes 120 L of water to produce one 1.3-L can of fruit juice and about 450 L of water to place one boiled egg on your breakfast plate.

What explanation can you give for these two facts?

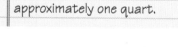

One liter (1 L) is a volume of approximately one quart.

As you just learned, one of the basic challenges that we face is related to the purity of the water we use. In the Laboratory Activity that follows, you will begin to address this challenge by trying to purify as much of a contaminated

Lab Video

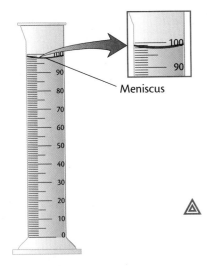

Figure 1 *To find the volume of liquid in a graduated cylinder, read the scale at the bottom of the curved part of the liquid (meniscus).*

Meniscus

The metric prefix *milli-* represents 1/1000th (0.001) of the unit indicated. Thus one milliliter (1 mL) is one-thousandth of a liter, or 0.001 L.

water ("foul water") sample as you can, producing water that is clean enough to use to rinse your hands. Even though most natural water samples are considerably less contaminated than your water sample, the challenge remains the same—to purify water supplies at low total cost to satisfy society's numerous needs.

A.2 FOUL WATER   **Laboratory Activity**

Introduction

Your objective is to purify a sample of "foul water," producing as much "clean water" as possible. **CAUTION:** *Do not test the purity of the water samples by drinking or tasting them.* You will use three water-purification procedures: (1) oil-water separation, (2) sand filtration, and (3) charcoal adsorption and filtration. In **filtration,** solid particles are sepa-rated from a liquid by passing the mixture through a material that retains the solid particles. The liquid collected after filtration is called the **filtrate.**

If you have not already done so, carefully read "Safety in the Laboratory," pages iv–v, before beginning the laboratory procedure.

Procedure

1. In your laboratory notebook, prepare a data table similar to the one shown here. Be sure to provide more space to write your entries.

2. Using a beaker, obtain approximately 100 mL (milliliters) of foul water from your teacher. Measure its volume precisely with a graduated cylinder. See Figure 1. Record the actual volume of the water sample in your data table.

3. Examine the properties of your sample: color, clarity, odor, and presence of oily or solid regions. Record your observations in the "Before treatment" row of your data table.

DATA TABLE

	Volume (mL)	Color	Clarity	Odor	Presence of Oil	Presence of Solids
Before treatment						
After oil-water separation						
After sand filtration						
After charcoal adsorption and filtration						

Oil-Water Separation

As you probably know, if oil and water are mixed and left undisturbed, the oil and water do not noticeably dissolve in each other. Instead, two layers form. Which layer floats on top of the other? Make careful observations in the following procedure to check your answer.

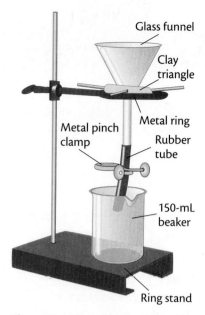

4. Place a funnel in a clay triangle supported by a ring and ring stand. See Figure 2. Attach a rubber tube to the funnel tip as shown.

5. Close the rubber tube by tightly pinching it with your fingers (or by using a metal pinch clamp). Gently swirl the foul-water sample for several seconds. Then immediately pour about half the sample into the funnel. Let it stand for a few seconds until the liquid layers separate. (Gentle tapping may encourage oil droplets to break free.)

6. Carefully open the tube, slowly releasing the lower liquid layer into an empty 150-mL beaker. Just as the lower layer has drained out, quickly close the rubber tube.

7. Drain the remaining layer into another 150-mL beaker.

8. Repeat Steps 5 through 7 for the other half of your sample, adding each liquid to the correct beaker. Which beaker contains the oily layer? How do you know?

9. Dispose of the oily layer as instructed by your teacher. Observe the properties of the remaining layer and measure its volume. Record your results. Save this water sample for the next procedure.

10. Wash the funnel with soap and water.

Figure 2 *Funnel in clay triangle.*

Sand Filtration

A **sand filter** traps and removes solid impurities—at least those particles too large to fit between sand grains—from a liquid.

11. Using a straightened paper clip, poke small holes in the bottom of a disposable cup. See Figure 3.

12. Add pre-moistened gravel and sand layers to the cup as shown in Figure 4. (The bottom layer of gravel prevents the sand from

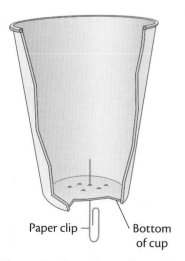

Figure 3 *Preparing a disposable cup.*

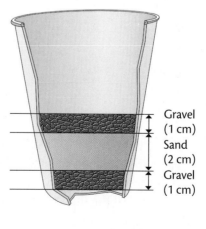

Gravel (1 cm)
Sand (2 cm)
Gravel (1 cm)

Figure 4 *Sand filtration.*

washing through the holes. The top layer of gravel keeps the sand from churning up when the water sample is poured into the cup.)

13. Gently pour the sample to be filtered into the cup. Catch the filtrate (filtered water) in a beaker as it drains through.

14. Dispose of the used sand and gravel according to your teacher's instructions. Do not pour any sand or gravel into the sink!

15. Observe the properties of the filtered water sample and measure its volume. Record your results. Save the filtered water sample for the next procedure.

Charcoal Adsorption and Filtration

Charcoal **adsorbs,** which means attracts and holds on its surface, many substances that could give water a bad taste, an odor, or a cloudy appearance. The pump system in a fish aquarium often includes a charcoal filter for this same purpose.

16. Fold a piece of filter paper as shown in Figure 5.

17. Place the folded filter paper in a funnel. Hold the filter paper in position and wet it slightly so that it rests firmly against the base and sides of the funnel cone.

18. Place the funnel in a clay triangle supported by a ring. See Figure 2, page 9. Lower the ring so that the funnel stem extends 2 to 3 cm (centimeters) inside a 150-mL beaker.

19. Place one teaspoon of charcoal in a 125- or 250-mL Erlenmeyer flask.

20. Pour the water sample into the flask. Swirl the flask vigorously for several seconds. Then gently pour the liquid through the filter paper. Keep the liquid level below the top of the filter paper—liquid should not flow through the space between the filter paper and the funnel. (Can you explain why?)

> There are about 2.5 cm (centimeters) in an inch. You can "think metric" and make a good estimation of the length of a centimeter by realizing that a cassette audio cartridge or a piece of chalk has a thickness of about 1 cm.

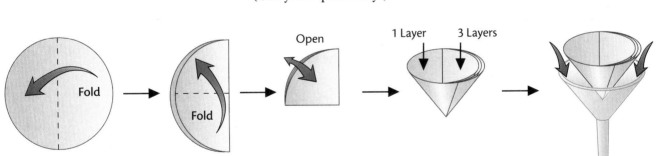

Figure 5 *Folding filter paper.*

21. If the filtrate is darkened by small charcoal particles, refilter the liquid through a clean piece of moistened filter paper.

22. When you are satisfied with the appearance and odor of your charcoal-filtered water sample, pour it into a graduated cylinder. Record the final volume and properties of the purified sample.

23. Follow your teacher's suggestions about saving your purified sample. Place used charcoal in the container provided by your teacher.

24. Wash your hands thoroughly before leaving the laboratory.

Calculations

Record all calculations and answers in your laboratory notebook.

1. What percent of your original "foul water" sample did you recover as "purified water"?

$$\text{Percent of water purified} = \frac{\text{Volume of water purified}}{\text{Volume of foul-water sample}} \times 100\%$$

2. What volume of liquid did you lose during purification?

3. What percent of your original foul-water sample was lost during purification?

Data Analysis

Prepare to answer the following questions by compiling a list of the "percent of water purified" (percent recovery) values for water samples from each laboratory group in your class.

1. Construct a **histogram** showing the percent recovery obtained by all laboratory groups in your class. To do so, organize the data into equal subdivisions such as 90.0–99.9%, 80.0–89.9%, and so forth. Count the number of data points in each subdivision. Then use this number to represent the height of the appropriate bar on your histogram, as illustrated in Figure 6. In this histogram you can see that three data points fell between 70.0% and 79.9% in this example.

2. What was the largest percent recovery obtained by a laboratory group in your class? The smallest? The difference between the largest and smallest values in a data set is the **range** of those data points. What is the range of percent recovery data in your class?

3. What was the average percent recovery for your class? The **average** value is computed by adding all the values together and dividing by the total number of values. The result is also called the **mean** value.

4. The mean is one measure of what statisticians call central tendency—an expression of the most "typical" or "representative" value in a data set. Another important measure of central tendency is the **median,** or middle value. To find the median of your percent recovery data, list all values in either ascending or descending order.

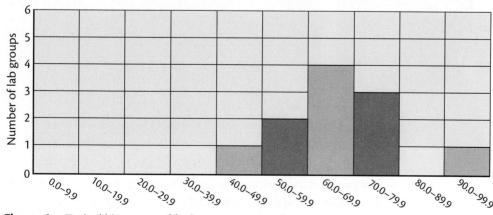

Figure 6 *Typical histogram of foul-water recovery data.*

Then find the value in the middle of the list—the point where there are as many data points above as below.

Consider this data set:

$$1 \quad 2 \quad 2 \quad 4 \quad 5 \quad 6 \quad 7$$

three values lower ↑ three values higher

median

If there is an even number of data points, take the average of the two values nearest to the middle.

What is the median percent recovery for your class?

Post-Lab Activities

Seeing & Imagining Water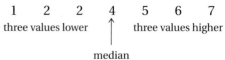

1. Your teacher will demonstrate **distillation,** another technique for water purification.
   a. Write a description of the steps in distillation.
   b. Why did your teacher discard the first portion of distilled liquid?
   c. Why was some liquid purposely left in the distilling flask at the end of the demonstration?

2. Your teacher will organize the testing of the **electrical conductivity** of purified water samples obtained by your class. This test focuses on the presence of dissolved, electrically charged particles in the water. You will also compare the electrical conductivity of distilled water and tap water. (You will read more about electrically charged particles on page 32.) What do these test results suggest about the purity of your water sample?

3. Your teacher will test the clarity of the various water samples by passing a beam of light through each sample. Observe the results. The differences are due to the presence or absence of the **Tyndall effect.** See Figure 7. What does this test suggest about the purity of your water sample?

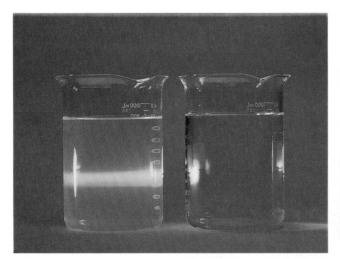 

Figure 7 *Particles are suspended in the sample in the beaker on the left. The particles are too small to be seen but large enough to reflect light. This is the Tyndall effect. Particles in the solution in the beaker on the right are too small to reflect light. The Tyndall effect is also observable in nature.*

Questions

1. Is your "purified water" sample "pure" water? How do you know?

2. Suggest how you might compare the quality of your water sample with that of other laboratory groups. That is, how can the relative success of each laboratory group be judged? Why?

3. How would you improve the water-purification procedures you followed so that a higher percent of purified water could be recovered?

4. a. Estimate the total time you expended in purifying your water sample.
   b. In your view, did that time investment result in a large enough sample of sufficiently purified water?
   c. In the real world, it is often said that "time is money."
      i. If you spent twice as much time in purifying your sample, would that extra investment in time "pay off" in higher-quality water?
      ii. If you spent about ten times as much time?
      Explain your reasoning.

5. Municipal water-treatment plants do not use distillation to purify the water. Why?

A.3 USES OF WATER

Making Decisions

Keep a diary of water use in your home for three days. On a data table similar to the one shown here, record how often various water-use activities occur. Ask each family member to cooperate and help you.

DATA TABLE

Per Household	Day 1	Day 2	Day 3
Number of persons			
Number of baths			
Number of showers Average duration of showers (min)			
Number of toilet flushes			
Number of hand-washed loads of dishes			
Number of machine-washed loads of dishes			
Number of washing-machine loads of laundry			
Number of lawn/garden waterings Average duration of waterings (min)			
Number of car washes			
Number of cups used for cooking and drinking			
Number of times water runs in sink Average duration of running (min)			
Other uses and frequency			

Check the activities listed on the chart. If family members use water in other ways in the three-day-period, add those uses to your diary. Estimate the quantities of water used by each activity. You may be surprised by the large amount of water that you and your family use. Perhaps you are inclined to wonder what characteristics make water such a widely used and important substance.

A.4 WATER SUPPLY AND DEMAND

A trillion gallons is 1 000 000 000 000 gallons, or 10^{12} gallons.

Is so much water used in the United States that the nation is in danger of running out? The answer is both no and yes. The total water available is far more than enough. Each day, some 15 trillion liters (4 trillion gallons) of rain or snow falls in the United States. Only 10% is used by humans. The rest flows back into large bodies of water, evaporates into the air, and falls again as part of a perpetual **water cycle,** or **hydrologic cycle.** So that is the "no" part of the answer. However, the distribution of rain and snow in the United States does not necessarily correspond to regions of high water use. Figure 8 summarizes how available water is used in various geographic regions of the country, organized by six major water-use categories.

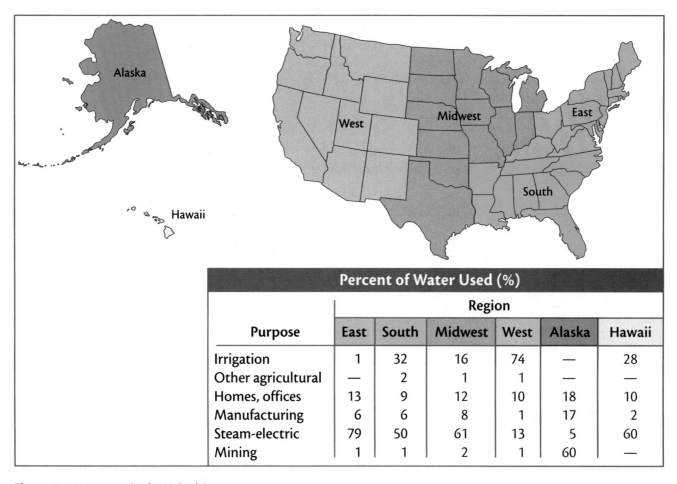

Percent of Water Used (%)

Purpose	East	South	Midwest	West	Alaska	Hawaii
Irrigation	1	32	16	74	—	28
Other agricultural	—	2	1	1	—	—
Homes, offices	13	9	12	10	18	10
Manufacturing	6	6	8	1	17	2
Steam-electric	79	50	61	13	5	60
Mining	1	1	2	1	60	—

Figure 8 *Water use in the United States.*

Refer to Figure 8 in answering these questions.

1. In the United States, what is the greatest single use of water in
   a. the East?
   d. the West?
   b. the South?
   e. Alaska?
   c. the Midwest?
   f. Hawaii?

2. Suggest reasons for differences in water-use priorities among these six U.S. regions.

3. Explain the differences in how water is used in the East and the West. Think about where most people live and where most of the nation's factories and farms are located. What other regional differences help explain patterns of water use?

An average family of four in the United States uses about 1200 liters (320 gallons) of water daily. That approximate volume represents direct, measurable water use. But in addition to that are many indirect, or hidden, uses of water that you probably have never considered. Each time you eat a slice of pizza, potato chips, or an egg, you are "using" water. Why? Because water was needed to grow and process the various components of each of these foods.

Consider again ChemQuandary 1 on page 7. At first glance, you probably thought that the volumes of water mentioned were absurdly large. How could so much water be needed to produce one boiled egg or one can of fruit juice? you might have asked. These two examples illustrate typical indirect (hidden) uses of water. The chicken that laid the egg needed drinking water. Water was used to grow the chicken's feed. Water was also used for various steps in the process that eventually brings the egg to your home. Even small quantities of water used for these and other purposes quickly add up when billions of eggs are involved!

In one of the Riverwood newspaper articles you read earlier, a youth was quoted as saying that he would drink bottled water and pre-packaged fruit juice until the water was turned back on. However, drinking fruit juice obtained from a grocery-store container involves the use of much more water than does drinking a glass of tap water. Why? Because the quantity of liquid in the container is insignificant when compared with the quantity of water used in making the container itself. The process of making a metal can, for example, is the source of the surprising 120 L of water mentioned on page 7! What examples of hidden water use associated with common materials do you encounter in daily life?

Although we depend on large quantities of water, we are relatively unaware of how much we use. This lack of awareness is understandable because water normally flows freely when taps are turned on—in Riverwood or in your home. Where does all this water come from? Check what you already know about sources of this plentiful supply.

A.5 WHERE IS EARTH'S WATER?

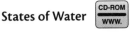

States of Water

You are probably not surprised to learn that most of Earth's total water (97% of it, in fact) is stored in the oceans. However, the next largest global water-storage place is not as obvious. Do you know what it is? If you said rivers and lakes, you and many others agree—however, your answer is incorrect. The second largest quantity of water is stored in Earth's ice caps and glaciers. Figure 9 shows how the world's supply of water is distributed.

As you know, water can be found in three different physical **states.** Water vapor in the air is in the **gaseous state.** Water is most easily identified in the **liquid state**—in lakes, rivers, oceans, clouds, and rain. Ice is a common example of water in the **solid state.** What other forms of "solid water" can you identify?

At present, most of the United States is fortunate to have abundant supplies of high-quality water. We turn on the tap, use what we need, and go about our daily routine—giving little thought to how that seemingly unlimited water supply manages to reach us.

If you live in a city or town, the water pipes in your home are linked to underground water pipes. These pipes bring water downhill from a reservoir or a water tower, usually located near the highest point in town, to all the faucets in the area. Water stored in the tower was cleaned and purified

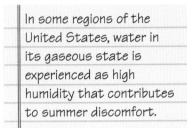

In some regions of the United States, water in its gaseous state is experienced as high humidity that contributes to summer discomfort.

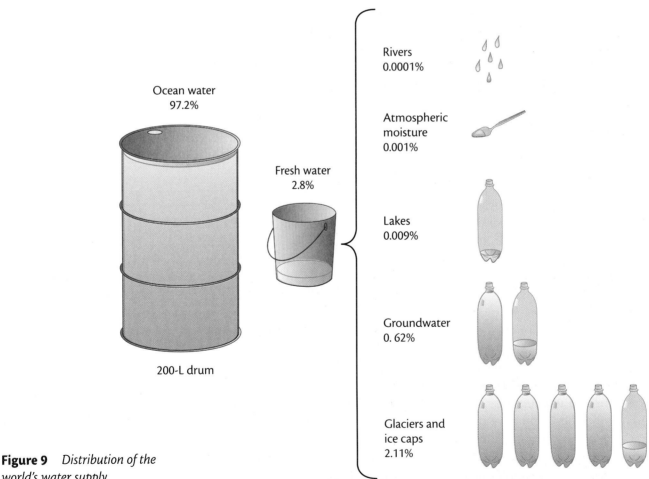

Figure 9 *Distribution of the world's water supply.*

at a water-treatment plant. It may have been pumped to the treatment plant from a reservoir, lake, or river. If your home's water supply originated in a river or other body of water, you are using **surface water.** If it originated in a well, you are using **groundwater.** Groundwater must be pumped to the surface.

If you live in a rural area, your home probably has its own water-supply system. A well with a pipe driven deep into an **aquifer** (a water-bearing layer of rock, sand, or gravel) pumps groundwater to the surface. A small pressurized tank holds the water before it enters your home's plumbing system.

Not surprisingly, neither groundwater nor surface-water samples are completely pure. When water falls as precipitation (rain, snow, sleet, or hail) and joins a stream or when it seeps far into the soil to become groundwater, it picks up small amounts of dissolved gases, soil, and rock. However, these dissolved substances are rarely removed from water at the treatment plant or from well water. In the amounts normally found in water, they are harmless. In fact, some minerals found in water (such as iron, zinc, or calcium) are essential in small quantities to human health or may improve water's flavor.

When the water supply is shut off, as it was in Riverwood, it is usually shut off for a short time. But suppose a drought lasted several years. Or suppose the shortage was perpetual, as it is in some areas of the world. In such circumstances, what uses of water would you give up first? It is clear that the use of available water for survival purposes would have priority. Nonessential uses would probably be eliminated.

Now it is time to examine your water-use data. Refer to your completed water-use diary to find out how much water you use in your home and for what purposes. The table in Figure 10 on page 18 lists typical quantities of water used in the home for common purposes.

> About one-fifth of the nation's water supply is groundwater.

A.6 WATER-USE ANALYSIS   Making Decisions

Use the table in Figure 10 on page 18 to answer the following questions regarding your household's water use, as well as those of your classmates.

Questions

1. Estimate the total water volume (in liters) used by your household during the three days.
2. How much water (in liters) did one member of your household use, on average, in one day?

Compile the answers to Question 2 for all members of your class by creating a histogram. To review the construction of a histogram, see page 11. (*Hint:* The range in value for each histogram bar must be equal and should be chosen so that there are about ten total bars.)

3. What is the range of average daily personal water use within your class?
4. Calculate the mean and median values for the class data. Which do you think is more representative of the data set? That is, which is a better expression of central tendency for these data?

> The table in Figure 10 indicates that a regular showerhead delivers 19 L each minute. In English units, that is 5 gallons each minute, or 25 gallons during a five-minute shower! To picture that volume of water, think of 25 gallons as 47 two-liter beverage bottles, or roughly twice the volume of gasoline in a normal automobile "fill up."

Figure 10 *Water required for typical activities.*

Water Required for Typical Activities	
Activity	**Volume of Water (L)**
Bathing (per bath)	130
Showering (per min)	
Regular showerhead	19
Water-efficient showerhead	9
Cooking and drinking	
Per 10 cups of water	2
Flushing toilet (per flush)	
Conventional toilet	19
"Water saver" toilet	13
"Low flow" toilet	6
Watering lawn (per hour)	1130
Washing clothes (per load)	170
Washing dishes (per load)	
By hand (with water running)	114
By hand (washing and rinsing in dishpans)	19
By machine (full cycle)	61
By machine (short cycle)	26
Washing car (running hose)	680
Running water in sink (per min)	
Conventional faucet	19
Water-saving faucet	9

5. Compare your answer to Question 2 with the estimated total volume of water, 300 L, used daily by each person in the United States. What reasons can you give to explain any difference between your value and the national value?

6. Which is closer to the national average (mean) for daily water use by each person, your answer to Question 2 or the class average in Question 4? What reasons can you give to explain why that value is closer?

EMERGENCY WATER FOR RIVERWOOD

ChemQuandary 2

Recall that the Riverwood town council arranged to truck water from Mapleton to Riverwood for three days to meet the needs of Riverwood residents for drinking water and cooking water. The current population of Riverwood is about 19 500.

1. On the basis of the water-use data collected and analyzed by your class (pages 13–14 and 17–18), explain how you might estimate the total volume (in liters or gallons) of water actually trucked to Riverwood during the three days.

2. What additional information would help you to improve your water-volume estimate? Why?

3. What assumptions must you make to complete your estimate?

You are now quite aware of the amount of water that you use daily. Suppose you had to live with much less. How would you ration your water allowance for survival and comfort? This is exactly the question that Riverwood residents are confronting.

A.7 RIVERWOOD WATER USE   Making Decisions

Riverwood authorities have severely rationed your home water supply for three days while possible fish-kill causes are investigated. The County Sanitation Commission recommends cleaning and rinsing your bathtub, adding a tight stopper, and filling the tub with water. That water will be your family's total water supply for every use other than drinking and cooking for the three-day period. (Recall that water for drinking and cooking will be trucked in from Mapleton.)

Assuming that your household has just one tub of water, 150 L (40 gal), to use during these three days and considering the typical water uses listed here, answer the questions below.

- washing cars, floors, windows, pets
- bathing, showering, washing hair, washing hands
- washing clothes, dishes
- watering indoor plants, outdoor plants, lawn
- flushing toilets

1. Which water uses could you do without completely? What would be the consequences?

2. For which tasks could you reduce your water use? How?

3. Impurities added by using water for one particular use may not prevent its reuse for other purposes. For example, you might decide to save hand-washing water and use it later to bathe your pet dog.

   a. For which activities could you use such impure water?

   b. From which prior uses could this water be taken?

Questions & Answers

It should now be obvious that clean water is a valuable resource that must not be taken for granted. See Figure 11. Unfortunately, water is easily contaminated. In the next section, as Riverwood works on dealing with its water emergency, you will examine some of the causes of water contamination.

Figure 11   *Water-conservation arithmetic*

SECTION SUMMARY

Reviewing the Concepts

♦ **Both direct and indirect uses of water must be considered when evaluating water use.**

1. Explain why placing a paper label on a juice container is an indirect use of water.

2. If you let the water run while brushing your teeth, explain how you are using water
   a. directly.
   b. indirectly.

3. Assume that Riverwood resident Jimmy Hendricks drank just pre-packaged fruit juice during the water shortage. Does that mean he did not use any water? Explain.

4. Listed below are some water uses associated with the foul-water laboratory activity. Classify each as either a direct or an indirect water use. Explain your answers.
   a. the manufacture of the filter paper
   b. the pre-moistening of the sand and gravel
   c. the use of water to cool the distillation apparatus

♦ **Water can be purified by techniques such as oil–water separation, filtration, adsorption, and distillation. The use of each technique depends on the contaminants present, the level of water purity desired, and the resources available for purification.**

5. Identify several techniques that can be used to purify water in a household. Describe the nature of each technique.

6. What kinds of material were removed from your foul-water sample in each step of that laboratory activity?

7. a. Would the water-purification procedures that you used in the foul-water laboratory activity make seawater suitable for drinking? Explain.
   b. What, if any, additional purification steps would be needed to make seawater suitable for drinking?

♦ **The amount of water available in Earth's hydrologic cycle is essentially fixed. However, the distribution of water is not always sufficient to meet local needs.**

8. Has the world's supply of water changed in volume in the past 100 years? The past million years? Explain.

9. In regard to Earth's water supply, rank the following sources in order of greatest to least total abundance: rivers, oceans, glaciers, water vapor.

10. Consider this familiar quotation: "Water, water, everywhere, Nor any drop to drink." Describe a situation in which this statement would, in fact, be true.

11. Diagram the major steps of the hydrologic cycle.

Connecting the Concepts

12. Explain why it might be possible that a molecule of water that you drank today was once swallowed by a dinosaur.

13. You are marooned on a sandy island surrounded by ocean water. A stagnant, murky looking pond contains the only available water on the island. In your survival kit, you have the following items:
    - one nylon jacket
    - one plastic cup
    - two plastic bags
    - one length of rubber tubing
    - one knife
    - one 1-L bottle of liquid bleach
    - one 5-L glass bottle
    - one bag of salted peanuts

    Describe a plan to produce drinkable water using only the items available.

14. What would happen to Earth's hydrologic cycle if the evaporation of water suddenly stopped?

15. Consider these foul-water purification procedures: oil–water separation, sand filtration, charcoal adsorption and filtration, and distillation. Which procedure would be least practical to use for purifying a city's water supply? Why?

16. A group of friends is planning a four-day backpacking hike. Some members of the group favor carrying their own bottled water; others wish to carry no water and instead buy bottled water in towns along the trail; still others want to purchase and carry portable filters for use with stream water.

    a. What are the advantages and disadvantages of each option?

    b. What additional information would the group need before making a decision?

Extending the Concepts

17. A politician guarantees that if she is elected, every household will have 100% pure water in every tap. Evaluate this promise and predict the likelihood of its success.

18. Find out how much charcoal is used in a fish aquarium filter and how fast water flows through the filter. Estimate the volume of water that can be filtered by a kilogram of charcoal. How much charcoal would be needed to filter the daily water supply for Riverwood, population 19 500?

19. One unique characteristic of water is that it is present in all three physical states (solid, liquid, and gas) in the range of temperatures found on our planet. How would the hydrologic cycle be different if this were not true?

20. Charcoal-filter materials are available in various sizes—from briquettes to fine powder. List the advantages and disadvantages of using either large charcoal pieces or small charcoal pieces for filtering.

MEETING RAISES FISH-KILL CONCERNS

BY CAROL SIMMONS
Riverwood News Staff Reporter

More than 300 concerned citizens, many "armed" with lists of questions, attended a Riverwood Town Hall public meeting last night to hear from the scientists investigating the Snake River fish kill.

Dr. Harold Schmidt, a scientist with the Environmental Protection Agency (EPA), expressed regret that the fishing tournament was canceled but strongly supported the town council's action, saying that it was the safest course in the long run. He reported that his laboratory is conducting further river-water tests.

Dr. Margaret Brooke, a State University water specialist, helped interpret information and answered questions. Local physician Dr. Jason Martingdale and Riverwood High School home economics teacher Alicia Green joined the speakers during the question-and-answer session that followed.

Dr. Brooke confirmed that preliminary water-sample analyses showed no likely cause for the fish kill. She reported that EPA chemists will collect water samples at hourly intervals today to look for any unusual fluctuations in dissolved-oxygen levels, as fish must take in adequate oxygen gas dissolved in water.

Concerning possible fish-kill causes, Dr. Brooke said that EPA scientists have been unable to identify any microorganisms in the fish that could have been responsible for their death. She concluded that "it must have been something dissolved or temporarily suspended in the water." If dissolved matter were involved, important considerations would include the relative amounts of various substances that can dissolve in water and the effect of water temperature on their solubility. She expressed confidence that further studies would shed more light.

Dr. Martingdale reassured citizens that "thus far, no illness reported by either physicians or the hospital can be linked to drinking water." Ms. Green offered water-conservation tips for housekeeping and cooking to make life easier for inconvenienced citizens. The information sheet that she distributed is available at Town Hall.

Mayor Edward Cisko confirmed that water supplies will again be trucked in from Mapleton today and expressed hope that the crisis will last no longer than three days.

Those attending the meeting appeared to accept the emergency situation with good spirits. "I'll never take my tap water for granted again," said Trudy Anderson, a Riverwood resident. "I thought scientists would have the answers," puzzled Robert Morgan, head of Morgan Enterprises. "They don't know either! There's certainly more involved in all this than I ever imagined."

A LOOK AT WATER AND ITS CONTAMINANTS

SECTION B

As the preceding article indicates, scientists attribute the cause of the fish kill to something dissolved or suspended in the Snake River. What might those substances be? How can the search for the cause be narrowed further? Knowledge of the properties of water (and of substances that might be found in it) will aid in this task. To understand these properties, you will be introduced to matter at the particulate level. You will also begin to learn the language of chemistry and to use it to communicate with your classmates as you investigate the fish kill.

B.1 PHYSICAL PROPERTIES OF WATER

Water is a common substance—so common that it is usually taken for granted. You drink it, wash with it, swim in it, and sometimes grumble when it falls from the sky. But are you aware that water is one of the rarest and most unusual substances in the universe? As planetary space probes have gathered data, scientists have learned that the great abundance of water on Earth is unmatched by any planet or moon in our solar system. Earth is usually half-enveloped by water-laden clouds, as you can see in Figure 12. In addition, more than 70% of Earth's surface is covered by oceans having an average depth of more than three kilometers (two miles).

> Kilo- (k) is the metric prefix meaning 1000. One kilometer (km) = 1000 meters (m).

Figure 12 *Earth as seen from space. What states of water can be observed in this winter scene?*

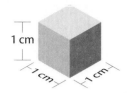

Figure 13 *One cubic centimeter (shown actual size).*

0 °C = 32 °F

Water is a form of matter. As you may recall from previous science courses, matter is anything that occupies space and has mass. All solids, liquids, and gases are classified as matter. Matter can be distinguished by its characteristic properties. Water has many important **physical properties,** properties that can be observed and measured without changing the chemical makeup of the substance. One physical property of matter is **density,** which is a measure of the mass of a material in a given volume. The density of water as a liquid is easy to remember. Because one milliliter (mL) of water has a mass of 1.00 g, the density of water is 1.00 g/mL. One milliliter of volume is exactly equal to one cubic centimeter (1 cm^3), which is pictured in Figure 13. Thus the density of water can also be reported as 1.00 g/cm^3. Another physical property of matter is freezing point. The freezing point of water is 0 °C at normal pressure. Can you think of other physical properties of water?

Water is the only ordinary liquid found naturally in our environment. Because so many substances dissolve readily in water, quite a few liquids are actually water solutions. Such water-based solutions are often called **aqueous solutions.** Even water that seems pure is never entirely so. Surface water contains dissolved minerals as well as other substances. Distilled water used in steam irons and car batteries contains dissolved gases from the atmosphere, as does rainwater.

Pure water is clear, colorless, odorless, and tasteless. The characteristic taste and slight odor of some tap-water samples are caused by substances dissolved in the water. You can confirm this by boiling and then refrigerating a sample of distilled water. When you compare its taste with the taste of chilled tap water, you may notice that "pure" distilled water tastes flat.

Water's physical properties, along with its chemical properties, distinguish it from other substances. In the following activity, you will compare the density of water with the density of some other common materials.

DENSITY   Building Skills 2

Most likely, you are already familiar with such physical properties of water as density, boiling point, and melting point. Use your experiences with water and other materials to answer the following questions concerning density.

1. In the foul-water laboratory activity (page 8), you observed that coffee grounds settled to the bottom of the water sample, whereas oil "floated" on top. Explain this observation in relation to the relative densities of coffee grounds, water, and oil.

2. How does the density of ice compare with that of liquid water? (*Hint:* Use your everyday experiences to answer this question.) What would happen to rivers and lakes (and fish) in cooler climates if the relative densities of ice and liquid water were reversed?

3. Suppose you were given a small cube of copper metal. What measurements would you need to make to determine its density? How would you make these measurements in the laboratory?

B.2 MIXTURES AND SOLUTIONS

How do you know when liquid water is not sufficiently pure? How can substances in water be separated and identified? Answers to these questions will be helpful in understanding and possibly solving the fish-kill mystery. But first you must learn how to recognize various types of mixtures.

When two or more substances combine yet retain their individual properties, the result is called a **mixture.** The foul water that you purified earlier is an example of a mixture because it contained coffee grounds, garlic powder, oil, and salt. As you discovered, the components of a mixture can be separated by physical means such as filtration and adsorption.

When you first examined your foul-water sample, did it look uniform throughout? Most likely, the coffee grounds had settled to the bottom and were not distributed evenly throughout the liquid. The foul water is an example of a **heterogeneous mixture** because its composition is not the same, or uniform, throughout. One type of heterogeneous mixture is called a **suspension** because the particles are large enough to settle out and can be separated by using a filter. Water plus coffee grounds and water plus small pepper particles are examples of suspensions.

If the particles are smaller than those in a suspension, they may not settle out and thus may cause the water to appear cloudy. Recall what happened when your teacher shined a light through your sample of purified water. The scattering of the light, known as the Tyndall effect (see Figure 7, page 12), indicated that small, solid particles were still present in the water. This type of mixture is known as a **colloid.**

A more familiar example of a colloid is milk, which contains small butterfat particles dispersed in water. These colloidal butterfat particles are not visible to the unaided eye; the mixture appears uniform throughout. Thus milk can be classified as homogeneous, which leads to the familiar term *homogenized milk.* Under high magnification, however, individual butterfat globules can be observed floating in the water. Milk no longer appears homogeneous. See Figure 14.

Particles smaller than colloidal particles also may be present in a mixture. When small amounts of table salt are mixed with water, as in your foul-water sample, the salt **dissolves** in the water. That is, the salt crystals separate into particles so small that they cannot be seen even at high

> A heterogeneous mixture's composition varies.

**Modeling Matter:
Attraction Between
Water Molecules**

Figure 14 *Fat globules can be seen under magnification, so that milk no longer looks homogeneous. Left: Whole milk under 1000X magnification. Center: Whole milk under 400X magnification. Right: Non-fat milk under 400X magnification.*

magnification. Nor do the particles exhibit the Tyndall effect when a light beam is passed through the mixture. These particles become uniformly mingled with the particles of water, producing a **homogeneous mixture,** or a mixture that is uniform throughout. All **solutions** are homogeneous mixtures. In a salt solution, the salt is the **solute** (the dissolved substance) and the water is the **solvent** (the dissolving agent). All solutions consist of one or more solutes and a solvent.

Evidence that something was still dissolved in your purified water sample came from the results of the conductivity test. The positive result (the bulb lit up) indicated that electrically charged particles were dissolved in the mixture.

Molecular Views of Water `CD-ROM` `WWW.`

> A compound can be broken down chemically into two or more simpler substances— either elements or new compounds. By definition, an element cannot be broken down into any simpler substances.

B.3 MOLECULAR VIEW OF WATER

So far in this investigation of water, you have focused on properties observable with your unaided senses. In doing so, have you wondered why water's freezing point is 0 °C or why certain substances such as salt dissolve in water? To understand why water has its particular properties, you must investigate it at the level of its atoms and molecules.

All matter is composed of **atoms.** Atoms are often called the building blocks of matter. Matter that is made up of only one kind of atom is known as an **element.** For example, oxygen is considered an element because it is composed of only oxygen atoms. Because hydrogen gas contains only hydrogen atoms, it too is an element. Approximately 90 different elements are found in nature, each having its own unique type of atom and identifying properties.

What type of matter is water? Is it an element? A mixture? As you most likely know, water contains atoms of two elements—oxygen and hydrogen. Thus water cannot be classified as an element. And, because its properties are different from those of oxygen and hydrogen, water cannot be classified as a mixture either.

Instead, water is an example of a **compound**—a substance composed of atoms of two or more elements linked together chemically in fixed proportions. To date, chemists have identified more than 18 million compounds. Compounds are represented by chemical formulas. In addition to water (H_2O), some other compounds and formulas with which you may be familiar include table salt (NaCl), ammonia (NH_3), baking soda ($NaHCO_3$), and chalk ($CaCO_3$).

Each element and compound is considered a **pure substance** because each has a uniform and definite composition as well as distinct properties. The smallest unit of a pure substance that retains the properties of that substance is a **molecule.** Atoms of a molecule are held together by **chemical bonds.** You can think of chemical bonds as the "glue" that holds atoms of a molecule together. One molecule of water is composed of two hydrogen atoms bonded to one oxygen atom, hence H_2O. An ammonia molecule (NH_3) contains three hydrogen atoms bonded to a nitrogen atom. Figure 15 shows representations of some atoms and molecules.

The following activity will give you a chance to apply an atomic and molecular view to a variety of common observations.

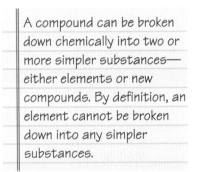

ATOMS   **MOLECULES**

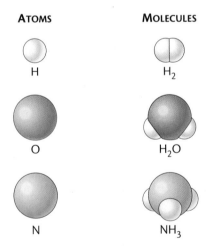

Figure 15   *On the left: hydrogen (H), oxygen (O), and nitrogen (N) atoms. On the right: hydrogen (H_2), water (H_2O), and ammonia (NH_3) molecules. Note the relative sizes of the atoms. Models similar to these are used throughout the textbook to depict atoms and molecules.*

PICTURES IN THE MIND

You live in a macroscopic world—a world filled with large-scale, readily observed things. As you experience the properties and behavior of bulk materials, you probably give little thought to the particulate world of atoms and molecules. If you wrap leftover cake in aluminum foil, it is unlikely that you think about how the individual aluminum atoms are arranged in the wrapping material. It is also unlikely that you consider what the mixture of molecules that make up air looks like as you breathe. And you seldom wonder about the behavior of atoms and molecules when you see water boiling or iron nails rusting.

Having a sense of how individual atoms and molecules might look and behave in elements, compounds, and mixtures can help you explain everyday phenomena. This activity will give you practice in observing, interpreting, evaluating, and creating visual models of matter at the particulate level.

To introduce you to these visualizations, consider this example: Suppose you want to draw a model of a homogeneous mixture of two gaseous compounds. You know that a homogeneous mixture is uniform throughout, so the two compounds should be intermingled and evenly distributed. You also know that compounds are composed of atoms of two or more different elements linked together by chemical bonds.

Suppose a molecule of one of the compounds contains two different atoms. To represent this molecule, you could draw two differently shaded or labeled circles to denote atoms of the two elements and a line connecting the atoms to denote a bond.

Suppose the other compound is composed of molecules that each contain three atoms, and that two of the atoms are of the same element. You now need to choose the order in which the atoms should be connected: the unique atom (Y) could be in the middle, X–Y–X, or on the end, X–X–Y. As long as you draw this imaginary compound in the same way every time, it does not matter which way you do it for this activity. However, the way in which atoms are connected in real compounds does, in fact, make a difference; X–Y–X is a different molecule from X–X–Y.

Examine the three models (a, b, and c) in the illustration. Which best represents a homogeneous mixture of the two compounds just described? You are correct if you said that b is the best visual model. The two types of molecules are uniformly mixed, and the atoms are shaded to indicate that they represent different elements. In a, the mixture is not homogeneous because the molecules are not uniformly mixed. Model c contains three different compounds instead of two. Notice that in a, bonded atoms in each molecule are connected by lines. In b and c, bonded atoms just touch each other. Both

a

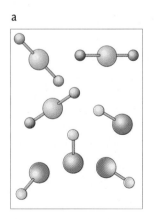

b

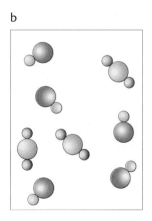

c

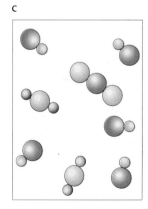

representations are used by chemists; either one is acceptable in this activity.

Now it is your turn to create and evaluate various visual models of matter.

1. Draw a model of a homogeneous mixture composed of three different gaseous elements. Describe the key features of your drawing.

2. What kind of matter does the following model represent? Explain your answer.

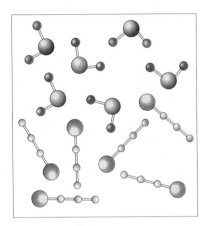

3. Draw a model of each of the following samples of matter. Write a description of key features of each model.

   a. a mixture of gaseous elements X and Z
   b. a two-atom compound of X and Z
   c. a four-atom compound of X and Z
   d. a solution composed of a solvent that is a two-atom compound of L and R, and a solute that is a compound composed of two atoms of D and one atom of T

4. One at a time, compare each visual model that you created in Question 3 with those of your classmates.

   a. Although the models may look a little different, does each set depict the same type of sample? Comment on any similarities and differences.
   b. Do the differences help or hinder your ability to visualize the type of matter being depicted?

5. The element iodine (I) has a greater density in the solid state than in the gaseous state. Draw models that depict and account for this difference at the atomic level. Iodine exists as a two-atom molecule.

6. A student in a chemistry class at Riverwood High School was asked to draw a model of a mixture composed of an element and a compound. Comment on the usefulness of the student's drawing.

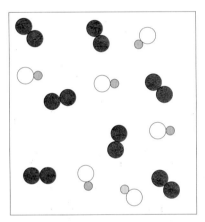

7. You have been interpreting and creating two-dimensional models of three-dimensional molecules.

   a. What are the limitations of two-dimensional representations?
   b. How can two-dimensional drawings be enhanced to show the features of three-dimensional atoms and molecules?
   c. Describe how you could make three-dimensional models from everyday materials.

8. a. How useful are models to you in visualizing matter at the particulate level?
   b. What characteristics do good models of matter have?

As you continue to study chemistry, you will encounter visual models of matter similar to those in this activity. When you see them, think about their usefulness as well as their possible limitations.

B.4 SYMBOLS, FORMULAS, AND EQUATIONS

An international "chemical language" for use in oral and written communication has been developed to represent atoms, elements, and compounds. The "letters" in this language's alphabet are **chemical symbols,** which are understood by scientists throughout the world. Each element is assigned a chemical symbol. Only the first letter of the symbol is capitalized; all other letters are lowercase. For example, C is the symbol for the element carbon and Ca is the symbol for the element calcium. Symbols for some common elements are listed in Figure 16.

All known elements are organized into the Periodic Table of the Elements, which is one of the most useful tools of chemists. As you continue your study of chemistry, you will learn more about this important table. For now, become familiar with this tool by locating each element listed in Figure 16 on the Periodic Table found on the inside back cover of this textbook. How many of these elements have you heard of before?

"Words" in the language of chemistry are composed of "letters" (elements) from the Periodic Table. Each "word" is a **chemical formula,** which represents a different chemical substance. In the chemical formula of a substance, a chemical symbol represents each element present. A **subscript** (a number written below the normal line of letters) indicates how many atoms of each element just to the left of the number are in one molecule or unit of the substance.

For example, as you already know, the chemical formula for water is H_2O. The subscript 2 indicates that each water molecule contains two hydrogen atoms. Each water molecule also contains one oxygen atom. However, the subscript 1 is understood and is therefore not included in a chemical formula. Here is another example. The chemical formula for propane, a compound commonly used as a fuel, is C_3H_8. What elements are present in propane, and how many atoms of each are there? You are correct if you said each propane molecule consists of three atoms of carbon and eight atoms of hydrogen.

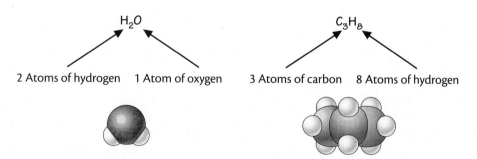

Common Elements	
Name	**Symbol**
Aluminum	Al
Bromine	Br
Calcium	Ca
Carbon	C
Chlorine	Cl
Cobalt	Co
Copper	Cu
Gold	Au
Hydrogen	H
Iodine	I
Iron	Fe
Lead	Pb
Magnesium	Mg
Mercury	Hg
Nickel	Ni
Nitrogen	N
Oxygen	O
Phosphorus	P
Potassium	K
Silver	Ag
Sodium	Na
Sulfur	S
Tin	Sn

Figure 16 *Common elements.*

If formulas are the "words" in the language of chemistry, then **chemical equations** are the "sentences." Each chemical equation summarizes the details of a particular chemical reaction. **Chemical reactions** entail the breaking and forming of chemical bonds, causing atoms to become rearranged into new substances. These new substances have different properties from those of the original material(s).

Elements That Exist as Diatomic Molecules	
Element	**Formula**
Hydrogen	H_2
Nitrogen	N_2
Oxygen	O_2
Fluorine	F_2
Chlorine	Cl_2
Bromine	Br_2
Iodine	I_2

Figure 17 *These elements occur naturally as diatomic molecules.*

"GEN-U-INE DIATOMICS" can serve as a good memory device for all common diatomic elements. The names of the diatomic elements end in either GEN or INE, and U better remember them!

The chemical equation for the formation of water

$$2\,H_2 \quad + \quad O_2 \quad \longrightarrow \quad 2\,H_2O$$

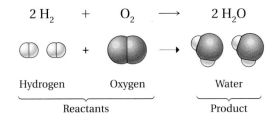

Hydrogen   Oxygen   Water

Reactants   Product

shows that two hydrogen molecules ($2\,H_2$) and one oxygen molecule (O_2) react to produce ($\rightarrow$) two molecules of water ($2\,H_2O$). The original (starting) substances in a chemical reaction are called the **reactants;** their formulas are always written on the left side of the arrow. The new substance or substances formed from the rearrangement of the reactant atoms are called **products;** their formulas are always written on the right side of the arrow. Note that this equation, like all chemical equations, is balanced—the total number of each type of atom (four H atoms and two O atoms) is the same for both reactants and products.

Perhaps you noticed that in the chemical equation for the formation of water the reactants hydrogen and oxygen are written with subscripts of 2 (H_2 and O_2). Do all elements have subscripts? Most uncombined elements in chemical equations are represented as single atoms (Cu, Fe, Na, and Mg, for example). A handful of elements are **diatomic molecules;** they exist as two bonded atoms of the same element. Oxygen and hydrogen are two examples of diatomic molecules. Figure 17 lists all the elements that exist as diatomic molecules at normal conditions. It will be helpful for you to remember these elements. Find the diatomic elements in the Periodic Table. Where are they located?

WORKING WITH SYMBOLS, FORMULAS, AND EQUATIONS

Building Skills 3

1. a. Name the element represented by each symbol below.
   - i. P
   - ii. Ni
   - iii. Cu
   - iv. Co
   - v. Br
   - vi. K
   - vii. Na
   - viii. Fe
   b. Which elements in Question 1a have symbols corresponding to their English names?
   c. Which is more likely to be the same throughout the world—the element's symbol or its name?

2. For each formula, name the elements present and give the number of atoms of each element.
   a. H_2O_2 Hydrogen peroxide (antiseptic)
   b. $CaCl_2$ Calcium chloride (de-icer for sidewalks)
   c. $NaHCO_3$ Sodium hydrogen carbonate (baking soda)
   d. H_2SO_4 Sulfuric acid (battery acid)

Look at the information available in a chemical equation:

$$N_2 \ + \ 3\,H_2 \ \longrightarrow \ 2\,NH_3$$

| Nitrogen gas | Hydrogen gas | Ammonia gas |

Household ammonia is made by dissolving gaseous ammonia in water.

First, complete an "atom inventory" of this chemical equation:

$$N_2 \ + \ 3\,H_2 \ \longrightarrow \ 2\,NH_3$$

2 N atoms + 6 H atoms = 2 N atoms and 6 H atoms

Note that the total number of atoms of N (nitrogen) and H (hydrogen) remains unchanged during this chemical reaction.

Next, interpret the equation in terms of molecules:

$$N_2 \ + \ 3\,H_2 \ \longrightarrow \ 2\,NH_3$$

1 N_2 molecule   3 H_2 molecules $\longrightarrow$ 2 NH_3 molecules

Note that one molecule of N_2 reacts with three molecules of H_2 to produce two molecules of the compound NH_3, called ammonia. Also note that molecules of nitrogen (N_2) and hydrogen (H_2) are diatomic, whereas the ammonia molecule is composed of four atoms—one nitrogen atom and three hydrogen atoms.

3. The following chemical equation represents the burning of methane, CH_4, to form water and carbon dioxide:

$$CH_4 + 2\,O_2 \longrightarrow CO_2 + 2\,H_2O$$

   a. Write a sentence describing the equation in terms of molecules.
   b. Identify each molecule as either a compound or an element.
   c. Complete an atom inventory for the equation.
   d. Provide a visual model ("picture in your mind") of the chemical reaction. Let [model] represent CH_4.

   Let [model] represent CO_2.

   Use the model of an H_2O molecule in Figure 15 (page 26) to draw a representation of H_2O similar to that of CH_4 and CO_2 shown here.

B.5 THE ELECTRICAL NATURE OF MATTER

Previously, you were introduced to the concept of atoms and molecules. How do the atoms in molecules "stick" together to form bonds? Are atoms made up of even smaller particles? The answers to these questions require an understanding of the electrical properties of matter.

   You have already experienced the electrical nature of matter, most probably without realizing it! Clothes often display "static cling" when they are taken from the dryer. The pieces of apparel stick firmly together and can be separated only with effort. The shock that you sometimes receive after walking across a rug and touching a metal doorknob is another reminder of matter's electrical nature. And if two inflated balloons are rubbed against your hair, both balloons will attract your hair but repel each other, a phenomenon best seen when the humidity is low.

The electrical properties of matter can be summarized as follows:

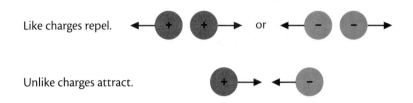

What are these positive and negative charges? How do they relate to the idea of atoms and molecules? The following points will be useful in answering these questions.

♦ Every electrically neutral (uncharged) atom contains equal numbers of positively charged particles called **protons** and negatively charged particles called **electrons.** In addition, most atoms contain one or more electrically neutral particles called **neutrons.**

♦ Positive–negative attractions between the protons in one atom and the electrons in another atom provide the "glue" that holds atoms together. This glue is the chemical bond that you read about on page 26.

States of Matter

These basic ideas will be used in later sections and in upcoming units to explain the properties of substances, the process of dissolving, and chemical bonding. Right now you will combine these ideas with your knowledge of atoms, chemical symbols, and chemical names to learn about a class of compounds that generally dissolve to some extent in water. It is possible that one or more of these compounds could be the cause of the fish kill.

B.6 IONS AND IONIC COMPOUNDS

Na	Electrically neutral sodium atom
Na⁺	Sodium ion
Cl	Electrically neutral chlorine atom
Cl⁻	Chloride ion
Na⁺Cl⁻	Sodium chloride (table salt)

Earlier in this unit (page 26), you learned about molecules. Molecules make up one type of compound. Another type of compound is composed of **ions,** which are charged atoms. Atoms can gain or lose electrons to form negative or positive ions, respectively. **Ionic compounds** are composed of positive and negative ions. An ionic compound has no net electrical charge; it is neutral because the positive and negative charges offset each other. The most familiar example of an ionic compound is table salt, sodium chloride (NaCl).

In solid ionic compounds, such as table salt, the ions are held together in crystals by attractions among the negative and positive charges. When an ionic compound dissolves in water, its individual ions separate from one another and disperse in the water. The designation (aq) following the symbol for an ion, as in Na^+(aq), means that the ions are in water (aqueous) solution.

When an atom gains one or more electrons (which have negative charge), the resulting negatively charged ion is called an **anion.** A positively charged ion, called a **cation,** results from an atom losing one or more electrons. An ion can be a single atom, such as a sodium ion (Na^+) or a chloride ion (Cl^-), or a group of bonded atoms, such as an ammonium ion (NH_4^+) or a nitrate ion (NO_3^-). An ion consisting of a group of bonded atoms is called a **polyatomic** (many-atom) **ion.** Figure 18 lists the formulas and names of common cations and anions.

Figure 18 *Common ions.*

Common Ions					
Cations					
1+ Charge		**2+ Charge**		**3+ Charge**	
Formula	Name	Formula	Name	Formula	Name
H^+	Hydrogen	Mg^{2+}	Magnesium	Al^{3+}	Aluminum
Na^+	Sodium	Ca^{2+}	Calcium	Fe^{3+}	Iron(III)*
K^+	Potassium	Ba^{2+}	Barium		
Cu^+	Copper(I)*	Zn^{2+}	Zinc		
Ag^+	Silver	Cd^{2+}	Cadmium		
NH_4^+	Ammonium	Hg^{2+}	Mercury(II)*		
		Cu^{2+}	Copper(II)*		
		Pb^{2+}	Lead(II)*		
		Fe^{2+}	Iron(II)*		
Anions					
1– Charge		**2– Charge**		**3– Charge**	
Formula	Name	Formula	Name	Formula	Name
F^-	Fluoride	O^{2-}	Oxide	PO_4^{3-}	Phosphate
Cl^-	Chloride	S^{2-}	Sulfide		
Br^-	Bromide	SO_4^{2-}	Sulfate		
I^-	Iodide	SO_3^{2-}	Sulfite		
NO_3^-	Nitrate	CO_3^{2-}	Carbonate		
NO_2^-	Nitrite				
OH^-	Hydroxide				
HCO_3^-	Hydrogen carbonate (bicarbonate)				

*Some metals form ions that have one charge under certain conditions and a different charge under different conditions. To specify the charge for these metal ions, Roman numerals are used in parentheses after the metal's name.

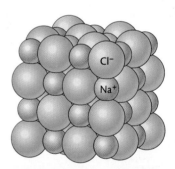

Figure 19 *Space-filling model of a sodium chloride (NaCl) crystal and a photo of magnified sodium chloride crystals.*

Solid sodium chloride, NaCl(s), consists of equal numbers of positive sodium ions (Na^+) and negative chloride ions (Cl^-) arranged in three-dimensional networks called crystals. See Figure 19. The ionic compound calcium chloride, $CaCl_2$, presents a similar picture. However, unlike sodium ions, calcium ions (Ca^{2+}) each have a charge of 2+.

You can easily write formulas for ionic compounds by following two simple rules.

1. Write the cation first, then the anion.

2. The correct formula contains the fewest positive and negative ions needed to make the total electrical charge zero.

Why are the numbers of chloride ions different in sodium chloride (NaCl) and calcium chloride ($CaCl_2$)? In sodium chloride, the ion charges are 1+ and 1−. Because one ion of each type results in a total charge of zero, the formula for sodium chloride must be NaCl.

When cation and anion charges do not add up to zero, ions of either type must be added until the charges cancel. In calcium chloride, one calcium ion (Ca^{2+}) has a charge of 2+. Each chloride ion (Cl^-) has a charge of 1−; two Cl^- ions are needed to equal a charge of 2−. Thus two chloride ions (2 Cl^-) are needed for each calcium ion (Ca^{2+}). The subscript 2 written after chlorine's symbol in the formula indicates this. The formula for calcium chloride is $CaCl_2$. Using these rules, what is the formula for aluminum sulfide, an ionic compound made up of aluminum cations (Al^{3+}) and sulfide anions (S^{2-})?

Formulas for compounds containing polyatomic ions, such as Na_2CO_3 (sodium carbonate), follow these same basic rules. However, if more than one polyatomic ion is needed to bring the total charge to zero, the formula for the polyatomic ion is enclosed in parentheses before the needed subscript is added. Ammonium sulfate is composed of ammonium (NH_4^+) and sulfate (SO_4^{2-}) ions. Two ammonium cations with a total charge of 2+ are needed to match the 2− charge of the sulfate anion. Thus the formula for ammonium sulfate is $(NH_4)_2SO_4$.

The written name of an ionic compound is composed of two parts. The cation is named first, then the anion. As Figure 18 (page 33) suggests, many cations have the same name as their original elements. Anions composed of a single atom, however, have the last few letters of the element's name changed to the suffix *-ide*. For example, the anion formed from fluorine (F) is fluor*ide* (F^-). Thus KF is named potassium fluoride. The following activity will provide practice in naming and writing formulas for ionic compounds according to the universal language of chemistry.

IONIC COMPOUNDS   Building Skills 4

Prepare a data table similar to the one shown here that identifies the composition of each ionic compound described in Statements 2 through 7.

DATA TABLE

	Cation	Anion	Formula	Name
1.	K^+	Cl^-	KCl	Potassium chloride
	(Complete this chart for substances 2 through 7.)			
7.				

Refer to Figure 18 on page 33 as needed to complete this activity. Potassium chloride, the primary ingredient in table-salt substitutes used by people on low-sodium diets, has been done as an example in the sample data table.

2. $CaSO_4$ is a component of plaster.

3. A substance composed of Ca^{2+} and PO_4^{3-} ions is found in some brands of phosphorus-containing fertilizer. This substance is also a major component of bones and teeth.

4. Ammonium nitrate, a rich source of nitrogen, is often used in fertilizer mixtures.

5. $Al_2(SO_4)_3$ is a compound that can be used to help purify water.

6. Magnesium hydroxide is called milk of magnesia when it is mixed with water.

7. Limestone and marble are two common forms of the compound calcium carbonate.

B.7 WATER TESTING

Laboratory Activity

Introduction

How can chemists detect and identify certain ions in water solutions? This activity will allow you to use a method that chemists, including those investigating the Riverwood fish kill, use to detect the presence of specific ions in water solutions.

The tests that you will perform in this activity are **confirming tests.** That is, a positive test confirms that the ion in question is present. In each confirming test, you will look for a change in solution color or for the appearance of an insoluble material called a **precipitate.** A negative test (no color or precipitate) does not necessarily mean that the ion in question is absent. The ion may simply be present in such a small amount that the test result may not be observed. Technologies are available to detect these very small amounts, however.

These tests are classified as qualitative tests, ones that identify the presence or absence of a particular substance in a sample. In contrast, quantitative tests determine the amount of a specific substance present in a sample. Both types of tests would most likely be used in determining the cause of the Snake River fish kill.

You will test for the presence of iron(III) (Fe^{3+}) and calcium (Ca^{2+}) cations, as well as chloride (Cl^-) and sulfate (SO_4^{2-}) anions. Although you are familiar with the names and symbols for Ca^{2+}, Cl^-, and SO_4^{2-}, the name for Fe^{3+} may look strange to you. Some elements can form cations with different charges. Iron atoms can lose either two electrons to form Fe^{2+} cations or three electrons to form Fe^{3+} cations. Thus the name "iron cation" is not descriptive enough; it does not distinguish between Fe^{2+} and Fe^{3+}. For this reason, Roman numerals are added to the name to indicate the charge on the ion. Examples of other cations that must include Roman numerals in their names are copper(I) and copper(II), and cobalt(II) and cobalt(III).

There are two types of iron cations: Fe^{2+} is designated Fe(II); Fe^{3+} is Fe(III).

You will perform each confirming test on several different water samples. The first solution will be a **reference solution**—one that contains the ion of interest. The second will be a **control**—a sample known not to contain the ion. The control in this activity is distilled water. The other solutions will be tap-water and natural-water samples that you or your teacher collected. These solutions may or may not contain the ion. To determine whether these solutions contain the ion, you will need to compare the results with your reference and control samples.

In your laboratory notebook, prepare four data tables (one for each ion) similar to the one shown. Add rows if you are testing more than one natural-water sample. Be certain to identify the source of each natural-water sample.

DATA TABLE: _____ (Specify ion)

Solution	Observations (color, precipitate, etc.)	Result (Is ion present?)
Reference		
Control		
Tap water		
Natural water from _____ (source)		

The following suggestions will help guide your ion analysis.

1. If the ion is in tap or natural water, it will probably be present in a smaller amount than in the same volume of reference solution. Thus the quantity of precipitate or color produced in the tap or natural water sample will be less than in the reference solution.

2. When completing an ion test, mix the contents of the well thoroughly, using a toothpick or small glass stirring rod. Do not use the same toothpick or stirring rod in other samples without first rinsing it and wiping it dry.

3. In a confirming test based on color change, so few color-producing ions may be present that it is difficult to determine if the reaction actually took place. Here are two ways to decide whether the expected color is actually present:

- Place a sheet of white paper behind or under the wellplate to make any color more visible.

- Compare the color of the control (distilled water) test with that of the sample. Distilled water does not contain any of the ions tested. So even a faint color in the tap or natural water confirms that the ion is present.

4. In a confirming test based on the formation of a precipitate, you may be uncertain whether a solid precipitate is present even after thoroughly mixing the solutions. Placing the wellplate on a black or dark surface often makes a precipitate more visible.

Procedure

The test procedures for each ion follow. If the ion of interest is present, a chemical reaction will take place, producing either a colored solution or a precipitate. The chemical equations are given for each ion.

> Only ions that take part in the reaction are included in this type of equation.

Calcium Ion (Ca²⁺) Test

$$Ca^{2+}(aq) \ + \ CO_3^{2-}(aq) \ \longrightarrow \ CaCO_3(s)$$

calcium ion   carbonate ion   calcium carbonate

Follow these steps for each sample (Ca²⁺ reference, control, tap water, natural water):

1. Place 20 drops into a well of a 24-well wellplate.
2. Add three drops of sodium carbonate (Na_2CO_3) test solution to the well.
3. Record your observations, including the color and whether a precipitate formed.
4. Determine whether the ion is present and record your results.
5. Repeat for the remaining solutions.
6. Discard the contents of the wellplate as directed by your teacher.

Iron(III) Ion (Fe³⁺) Test

$$Fe^{3+}(aq) \ + \ SCN^-(aq) \ \longrightarrow \ [FeSCN]^{2+}(aq)$$

iron(III) ion   thiocyanate ion   iron(III) thiocyanate ion

Follow these steps for each sample (Fe³⁺ reference, control, tap water, natural water):

1. Place 20 drops into a well of a 24-well wellplate.
2. Add one or two drops of potassium thiocyanate (KSCN) test solution to the well.
3. Record your observations, including the color and whether a precipitate formed.
4. Determine whether the ion is present and record your results.
5. Repeat for the remaining solutions.
6. Discard the contents of the wellplate as directed by your teacher.

Chloride Ion (Cl⁻) Test

$$Cl^-(aq) \ + \ Ag^+(aq) \ \longrightarrow \ AgCl(s)$$

chloride ion   silver ion   silver chloride

Follow the same procedure as that for the Fe³⁺ ion, with the following changes:

- Use the Cl⁻ reference solution.
- In Step 2, add three drops of silver nitrate ($AgNO_3$) test solution instead of potassium thiocyanate (KSCN) test solution.

Sulfate Ion (SO_4^{2-}) Test

$$SO_4^{2-}(aq) + Ba^{2+}(aq) \longrightarrow BaSO_4(s)$$

sulfate ion   barium ion   barium sulfate

Follow the same procedure as that for the Fe^{3+} ion, with the following changes:

- Use the SO_4^{2-} reference solution.
- In Step 2, add three drops of barium chloride ($BaCl_2$) test solution instead of potassium thiocyanate (KSCN).

Questions

1. a. Why was a control used in each test?
   b. Why was distilled water chosen as the control?
2. Describe some difficulties associated with the use of qualitative tests.
3. These tests cannot absolutely confirm the absence of an ion. Why?
4. How might your observations have changed if you had not cleaned your wells or stirring rods thoroughly between each test?

B.8 PURE AND IMPURE WATER

Now that you have learned about water's properties and about some of the substances that can dissolve in water, you are ready to return to the problem of Riverwood's fish kill. Recall that various Riverwood residents had different ideas about the cause of the problem. For example, longtime resident Harmon Lewis was sure the cause was pollution of the river water. Which substances are regarded as pollutants, and which are harmless when dissolved in water?

Families in most U.S. cities and towns receive an abundant supply of clean, but not absolutely pure, water at an extremely low cost. You can check the water cost in your own area: If you use municipal water, your family's water bill will contain the current water cost per gallon. Divide that value by 3.8 (there are 3.8 liters in one gallon) to compute the current cost for one liter of water.

What is the difference between clean and pure water?

It is useless to insist on absolutely pure water. The cost of processing water to make it completely pure would be prohibitively high. And, even if costs were not a problem, it would still be impossible to have absolutely pure water. The atmospheric gases nitrogen (N_2), oxygen (O_2), and carbon dioxide (CO_2) will always dissolve in the water to some extent.

B.9 THE RIVERWOOD WATER MYSTERY

Making Decisions

Your teacher will divide the class into several different groups of students. Each group will complete this decision-making activity. Afterward, the entire class will compare and discuss the answers obtained by each group.

At the beginning of this unit, you read newspaper articles describing the Riverwood fish kill and the reactions of several citizens to it. Among those interviewed were Harmon Lewis, a longtime resident of Riverwood, and Dr. Margaret Brooke, a water-systems scientist. These two people had very different reactions to the fish kill. An angry Harmon Lewis was certain that human activity—probably some sort of pollution—had caused the fish kill. Dr. Brooke refused to even speculate about the cause of the fish kill until she had conducted some tests.

Which of these two positions comes closer to your own reaction at this point? Complete the following activities to investigate the issue further.

1. Reread the fish-kill newspaper reports located at the beginning of Sections A and B. List all facts (not opinions) concerning the fish kill found in these articles. Scientists often refer to facts as data. **Data** are objective pieces of information. They do not include interpretation.

2. List at least five factual questions that you would want answered before you could decide on possible causes of the fish kill. Some typical questions might be: Do barges or commercial boats travel on the Snake River? Were any shipping accidents on the river reported recently?

3. Look over your two lists—one of facts and the other of questions.

    a. At this point, which possible fish-kill causes can you rule out as unlikely? Why?

    b. Can you suggest a probable cause? Be as specific as possible.

Later in this unit you will have an opportunity to test the reasoning that you used in answering these questions.

B.10 WHAT ARE THE POSSIBILITIES?

The activities that you just completed (gathering data, seeking patterns or regularities among the data, suggesting possible explanations or reasons to account for the data) are typical of the approaches scientists take in attempting to solve problems. Such scientific methods are a combination of systematic, step-by-step procedures and logic, as well as occasional hunches and guesses.

A fundamental yet difficult part of scientists' work is knowing what questions to ask. You have listed some questions that might be posed concerning the cause of the fish kill. Such questions help focus a scientist's thinking. Often a large problem can be reduced to several smaller problems or questions, each of which is more easily managed and solved.

The number of possible causes for the fish kill is large. Scientists investigating this problem must find ways to eliminate some causes and zero in on more promising ones. They try to either disprove all but one cause or produce conclusive proof in support of a specific cause.

As you recall, water analyst Brooke studied possible causes of the Snake River fish kill. She concluded that if the actual cause were water related, it would have to be due to something dissolved or suspended in the water.

 Questions & Answers

In Section C, you will examine several categories of water-soluble substances and consider how they might be implicated in the fish kill. The mystery of the Riverwood fish kill will be confronted at last!

Environmental Cleanup: It's a Dirty Job . . . But That's the Point

Wayne Crayton spends his summers touring exotic islands in the Aleutian Islands chain off the coast of Alaska. But it's not just an adventure that he's embarked on. It's also his job.

As an Environmental Contaminants Specialist with the United States Army Corps of Engineers, Alaska District, Wayne investigates areas that were formerly used as military bases and fueling stations. Wayne and his teammates review and assess the damage (if there is any) that contaminants have done to key areas used by wildlife. Based on their findings, they then develop plans to fix the problems.

> Wayne and his teammates recommend procedures for removing and treating contaminated soil and other material.

As part of its investigation planning, Wayne's team reviews information to determine what they're likely to find at a given site. For example, historical documents about the site will indicate whether the team members should be looking for petroleum residues or other contaminants; aerial photography and records from earlier investigations will help to identify specific areas that are potential sites of contamination.

At the site, the team collects soil, sediment, and water samples from the exact location where a contaminant was originally introduced to the environment, as well as samples from the area over which the contaminant might have spread. Wayne and his teammates may also collect small mammals or fish that have been exposed to the contaminants. After having collected the necessary samples, the team members return home quickly because some of the collected samples can degrade or change characteristics soon after collection.

Wayne works in an office in Anchorage during the rest of the year, analyzing and interpreting data and test results from the field investigations. He and his colleagues calculate concentrations of hazardous substances, including organochlorines, PCBs (polychlorinated biphenyls), pesticides, petroleum residues, and trace metals. Then they determine whether any of these substances present a risk to humans or the surrounding ecosystems.

Using these results, Wayne and his teammates recommend procedures for removing and treating contaminated soil and other material. In some situations, they decide that the best solution is to do nothing; the cleanup itself could destroy wetlands, disturb endangered wildlife, or have other negative effects on the environment. The Corps of Engineers uses the team's recommendations to direct the work of the contractor performing the actual cleanup.

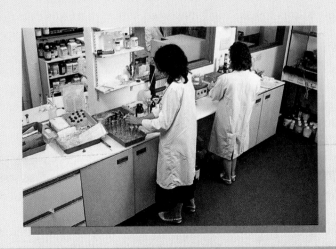

Solving Scientific Problems . . .

This icon indicates an opportunity to consult resources on the World Wide Web. See your teacher for further instructions.

Scientists often solve problems in unique ways—ways that are different from the methods used in other areas of academic research.

◆ Outline the problem-solving steps that the Environmental Contaminants Specialists use in planning their investigations as described in this article.

◆ Compare the steps used by these scientists with the steps that you have used in studying science.

◆ Conduct a World Wide Web search for any United States Army Corps of Engineers or United States Environmental Protection Agency investigations or projects that might be underway in your community.

SECTION SUMMARY

Reviewing the Concepts

♦ **Physically combining two or more substances produces a mixture. Mixtures are considered heterogeneous or homogeneous, depending on the distribution of materials in the mixture.**

1. When gasoline and water mix, they form two distinct layers. What do you need to know in order to determine which liquid will be found in the top layer?

2. Identify each of the following materials as a solution, suspension, or colloid. Explain your choice in each case.

   a. a medicine accompanied by the instructions "shake before using"
   b. Italian salad dressing
   c. mayonnaise
   d. a cola soft drink
   e. an oil-based paint
   f. milk

3. You notice beams of light passing into a darkened room through blinds on a window. Does this demonstrate that the room air is a solution, suspension, or colloid? Explain.

4. Sketch a visual model on the molecular level that represents each of the following types of mixtures. Label and explain the features of each sketch.

   a. a solution   c. a colloid
   b. a suspension

5. a. Given a mixture, what steps would you follow to classify it as a solution, a suspension, or a colloid?

   b. Describe how each step would help you to distinguish among the three types of mixtures.

♦ **All matter is composed of atoms. An element is composed of only one type of atom; compounds consist of two or more types of atoms. Elements and compounds are considered pure substances, each having unique physical and chemical properties.**

6. Using your knowledge of chemical symbols, classify each of the following substances as an element or a compound.

   a. CO   c. HCl   e. $NaHCO_3$ g. I_2
   b. Co   d. Mg   f. NO

7. Compare the physical properties of water (H_2O) with the physical properties of the elements of which it is composed.

8. Look at the following drawings.

   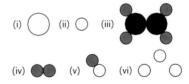

   a. Which represent a pure element?
   b. Which represent a compound?

♦ **A chemical formula for a substance contains chemical symbols and subscripts (if needed) that identify the type and number of atoms present in one molecule or unit. A chemical equation states how a substance or substances react to form new substances.**

9. Represent each chemical equation with drawings of the molecules and their component atoms. Use circles of different sizes or shading for each type of element.

   a. $H_2(g) + Cl_2(g) \longrightarrow 2\ HCl(g)$
   b. $2\ H_2O_2(aq) \longrightarrow 2\ H_2O(l) + O_2(g)$
      Let ⦿●● represent a hydrogen peroxide molecule, H_2O_2.

   c. Using complete sentences, write a word equation for the chemical equations given in a and b. Include the numbers of molecules.

10. Name the elements and list the number of atoms of each for the following substances.

   a. phosphoric acid, H_3PO_4 (used in some soft drinks and to produce some fertilizers)

b. sodium hydroxide, NaOH (found in some drain cleaners)

c. sulfur dioxide, SO_2 (a by-product of burning most, if not all, types of coal)

11. Write chemical equations that represent the following word equations:

a. Baking soda ($NaHCO_3$) reacts with hydrochloric acid (HCl) to produce sodium chloride, water, and carbon dioxide.

b. During respiration, one molecule of glucose, $C_6H_{12}O_6$, combines with six molecules of oxygen to produce six molecules of carbon dioxide and six molecules of water.

♦ **An atom is composed of smaller particles (protons, neutrons, and electrons), each possessing a characteristic mass and charge. An electrically neutral atom has an equal number of protons and electrons.**

12. For each of the following elements, identify the number of protons or electrons needed for an electrically neutral atom.

a. Carbon: 6 protons __ electrons
b. Aluminum. __ protons 13 electrons
c. Lead: 82 protons __ electrons
d. Chlorine: __ protons 17 electrons

13. Decide whether each of the following atoms is electrically neutral.

a. Oxygen: 16 protons 18 electrons
b. Iron: 26 protons 24 electrons
c. Silver: 47 protons 47 electrons
d. Iodine: 53 protons 54 electrons

♦ **Ionic compounds are composed of equal numbers of positively and negatively charged ions (atoms that have lost or gained electrons), thus giving the compound no net charge.**

14. Write the symbol and show the charge (if any) on the following atoms or ions:

a. hydrogen with 1 proton and 1 electron
b. sodium with 11 protons and 10 electrons
c. chlorine with 17 protons and 18 electrons
d. aluminum with 13 protons and 10 electrons

15. Indicate whether an Fe^{3+} ion would be attracted to or repelled from each particle in Question 14.

16. a. Classify each of the following as an electrically neutral atom, an anion, or a cation.

i. O^{2-} iii. C   v. Hg^{2+}
ii. He   iv. Ag^+

b. For each ion, indicate whether the electrical charge resulted from an atom gaining electrons, losing electrons, or neither.

17. Write the name and formula for each compound that will be formed from the following combinations of cations and anions:

	OH^-	$PO_4{}^{3-}$	S^{2-}
Fe^{3+}	a.	b.	c.
K^+	d.	e.	f.
Ca^{2+}	g.	h.	i.

Connecting the Concepts

18. Explain the possible risks in failing to follow the direction "Shake before using" on the label of a medicine bottle.

19. Why is it important that the symbols of the elements be internationally accepted?

20. Draw a model of a solution in which water is the solvent and oxygen gas (O_2) is the solute.

21. An iron atom that has 26 protons and 23 electrons combines with an O^{2-} ion to form a compound.

a. What is the ionic charge on the iron atom?
b. Write the chemical formula for the compound.

Extending the Concepts

22. Is it possible to have a food product that is 100% "chemical free"? Explain.

23. Some elements in Figure 16 (page 29) have symbols that are not based on their modern names (such as K for potassium). Look up their historical names and explain the origin of their symbols.

24. The symbols of elements (such as Na, Cu, and Cl) are accepted and used by chemists in all nations, regardless of the country's official language. However, the name of an element often depends on language. For example, the element N is "nitrogen" in English but "azote" in French. The element H is "hydrogen" in English but "Wasserstoff" in German. Investigate the names of some common elements in a foreign language of your choice. What are the meanings or origins of the foreign element names that you have found? How do those meanings or origins compare with those for the corresponding English element names?

25. Investigate and report on why "100% pure water" would be unsuitable for long-term human consumption—even if taste were not a consideration.

26. Using an encyclopedia or other reference, compare the maximum and minimum temperatures naturally found on the surfaces of Earth, the Moon, and Venus. The large amount of water on Earth serves to limit the natural temperature range on the planet. Suggest ways that water accomplishes this. As a start, find out what *heat of fusion, heat capacity,* and *heat of vaporization* mean.

27. Look up the normal freezing point, boiling point, heat of fusion, and heat of vaporization of ammonia (NH_3). If a planet's life forms were made up mostly of ammonia rather than water, what special survival problems might they face? What temperature range would an ammonia-based planet need to support "life"?

SECTION C

INVESTIGATING THE CAUSE OF THE FISH KILL

The challenge facing investigators of the Riverwood fish kill is to decide what in Snake River water was responsible for the crisis. In this section, you will learn about the process of dissolving, how solutions behave and are described, and what types of substances dissolve in water. What you learn will help to ensure that you have the knowledge and skills needed to evaluate the Riverwood data and to determine the cause of the fish kill.

C.1 SOLUBILITY OF SOLIDS

Could something dissolved in the Snake River have caused the fish kill? As you already know, a variety of substances can dissolve in natural waters. To determine whether any of these substances could be harmful to fish, you first need to know how solutions are formed and described. For example, how much of a certain solid substance will dissolve in a given amount of water?

Imagine preparing a water solution of potassium nitrate, KNO_3. What happens as you add a scoopful of solid, white potassium nitrate crystals to water in a beaker? As you stir the water, the solid crystals dissolve and disappear. The resulting solution remains colorless and clear. In this solution, water is the solvent, and potassium nitrate is the solute.

What will happen if you add a second scoopful of potassium nitrate crystals to the beaker and stir? These crystals also may dissolve. However, if you continue adding potassium nitrate without adding more water, eventually some potassium nitrate crystals will remain undissolved on the bottom of the beaker, no matter how long you stir. The maximum quantity of a substance that will dissolve in a certain quantity of water (for example, 100 g) at a specified temperature is the **solubility** of that substance in water. In this example, the solubility of potassium nitrate might be expressed as "grams potassium nitrate per 100 g water" at a specified temperature.

From everyday experiences, you probably know that both the size of the solute crystals and the vigor and duration of stirring affect how long it takes for a sample of solute to dissolve at a given temperature. But do these factors affect how much substance will eventually dissolve? With enough time and stirring, will even more potassium nitrate dissolve in water? It turns out that the solubility of a substance in water is a characteristic of the substance and cannot be changed by any amount of stirring or time.

So what does affect the actual quantity of solute that dissolves in a given amount of solvent? As you can see from Figure 20 on page 46, the mass of

> Remember: In a solution, the solvent is the dissolving agent and the solute is the dissolved substance.

Questions & Answers

 Dissolving Ionic Compounds

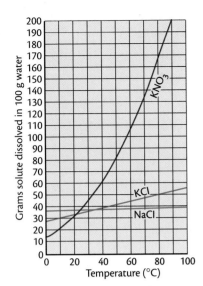

Figure 20 *Relationship between solute solubility in water and temperature.*

solid solute that will dissolve in 100 g water varies as the temperature of the water changes from 0 °C to 100 °C. The graphical representation of this relationship is called the solute's solubility curve.

Each point on the solubility curve indicates a solution in which the solvent contains as much dissolved solute as it normally can at that temperature. Such a solution is called a **saturated solution.** Thus each point on the solubility curve indicates a saturated solution. Look at the curve for potassium nitrate (KNO_3) in Figure 20. At 50 °C, how much potassium nitrate will dissolve in 100 g water to form a saturated solution? This value—80 g KNO_3 per 100 g water—is the solubility of potassium nitrate in 50 °C water. In contrast, the solubility of potassium nitrate in 20 °C water is only about 30 g KNO_3 per 100 g water. (Make sure that you are able to "read" this value on the graph.)

Note that the solubility curve for sodium chloride (NaCl) is nearly a horizontal line. What do you think this means about the solubility of sodium chloride as temperature changes? Compare the curve for sodium chloride with the curve for potassium nitrate (KNO_3), which rises steeply as temperature increases. You should be able to conclude that for some solutes, such as potassium nitrate (KNO_3), solubility in water is greatly affected by temperature, whereas for others, such as sodium chloride (NaCl), the change is only slight.

Now consider a solution containing 80 g potassium nitrate in 100 g water at 60 °C. Locate this point on the graph. Where does it fall with respect to the solubility curve? What does this tell you about the level of saturation of the solution? Because each point on the solubility curve represents a saturated solution, any point on a graph below a solubility curve must represent an unsaturated solution. An **unsaturated solution** is a solution that contains less dissolved solute than the solvent can normally hold at that temperature.

What would happen if you cooled this solution to 40 °C? (Follow the line representing 80 g to the left on the graph.) You might expect that some solid KNO_3 crystals would form and fall to the bottom of the beaker. In fact, this event is likely to occur. Sometimes, however, you can cool a saturated solution without forming any solid crystals, producing a solution that contains more solute than could usually be dissolved at that temperature. This type of solution is called a **supersaturated solution.** (Note that this new point lies above the solubility curve for potassium nitrate.) Agitating a supersaturated solution or adding a "seed" crystal to the solution often causes the "extra" solute to appear as solid crystals and settle to the bottom of the beaker, or precipitate. The remaining liquid then contains the amount of solute that represents a stable, saturated solution at that temperature.

One example of crystallization in a supersaturated solution may be familiar to you—the production of rock candy. A water solution is supersaturated with sugar. When seed crystals are added, they cause excess dissolved sugar to crystallize from the solution onto a string. Mineral deposits around a hot spring are another example of crystallization from a supersaturated solution. Water emerging from a hot spring is saturated with dissolved minerals. As the solution cools, it becomes supersaturated. The rocks over which the solution flows act as seed crystals, causing the formation of more mineral deposits.

SOLUBILITY AND SOLUBILITY CURVES

What is the solubility of potassium nitrate at 40 °C? The answer is found by using the solubility curve for potassium nitrate given in Figure 20. Locate the intersection of the potassium nitrate curve with the vertical line representing 40 °C. Follow the horizontal line to the left and read the value. The solubility of potassium nitrate in water at 40 °C is 60 g per 100 g water.

At what temperature will the solubility of potassium chloride be 25 g per 100 g water? Think of the space between 20 g and 30 g on the y axis in Figure 20 as divided into two equal parts, then follow an imaginary horizontal line at "25 g/100 g" to its intersection with the curve. Follow a vertical line down to the x axis. Because the line falls halfway between 10 °C and 20 °C, the desired temperature must be about 15 °C.

As you have seen, the solubility curve is quite useful when you are working with 100 g water. But what happens when you are working with other quantities of water? The solubility curve indicated that 60 g potassium nitrate will dissolve in 100 g water at 40 °C. How much potassium nitrate will dissolve in 150 g water at this temperature? You can "reason" the answer in the following way.

The amount of solvent (water) has increased from 100 g to 150 g—1.5 times as much solvent. That means that 1.5 times as much solute can be dissolved. Thus: $1.5 \times 60 \text{ g} = 90 \text{ g KNO}_3$.

The calculation can also be written as a simple proportion, which will give the same answer:

$$\frac{60 \text{ g KNO}_3}{100 \text{ g H}_2\text{O}} = \frac{x \text{ g KNO}_3}{150 \text{ g H}_2\text{O}}$$

$$x \text{ g KNO}_3 = \frac{(60 \text{ g KNO}_3)(150 \text{ g H}_2\text{O})}{(100 \text{ g H}_2\text{O})} = 90 \text{ g KNO}_3$$

Refer to Figure 20 to answer the following questions.

1. a. What mass (in grams) of potassium nitrate (KNO_3) will dissolve in 100 g water at 60 °C?
   b. What mass (in grams) of potassium chloride (KCl) will dissolve in 100 g water at this temperature?

2. a. You dissolve 25 g potassium nitrate in 100 g water at 30 °C, producing an unsaturated solution. How much more potassium nitrate (in grams) must be added to form a saturated solution at 30 °C?
   b. What is the minimum mass (in grams) of 30 °C water needed to dissolve 25 g potassium nitrate?

3. a. A supersaturated solution of potassium nitrate is formed by adding 150 g KNO_3 to 100 g water, heating until the solute completely dissolves and then cooling the solution to 55 °C. If the solution is agitated, how much potassium nitrate will precipitate?
   b. How much 55 °C water would have to be added (to the original 100 g water) to just dissolve all of the KNO_3?

C.2 SOLUTION CONCENTRATION

The general terms saturated and unsaturated are not always adequate for describing the properties of solutions that contain different amounts of solute. A more precise description of the amount of solute in a solution is needed—an exact, numerical measure of concentration.

Solution concentration refers to the quantity of solute dissolved in a specific quantity of solvent or solution. You have already worked with one type of solution concentration expression: The water-solubility curves in Figure 20 (page 46) reported solution concentrations as the mass of a substance dissolved in a given mass of water.

Another way to express concentration is with percents. For example, dissolving 5 g table salt in 95 g water produces 100 g solution with a 5% salt concentration (by mass).

$$\frac{5 \text{ g salt}}{100 \text{ g solution}} \times 100\% = 5\% \text{ salt solution}$$

"Percent" means parts per hundred parts. So a 5% salt solution could also be reported as five parts per hundred of salt (5 pph salt). However, percent is much more commonly used.

For solutions containing much smaller quantities of solute (as are found in many environmental water samples, including those from the Snake River), concentration units of **parts per million (ppm)** are sometimes useful. What is the concentration of the 5% salt solution expressed in ppm? Because 5% of 1 million is 50 000, a 5% salt solution is 50 000 parts per million.

Although you may not have realized it, the notion of concentration is part of daily life. For example, preparing beverages from concentrates, adding antifreeze to an automobile, and mixing pesticide or fertilizer solutions all require the use of solution concentrations. The following activity will help you review the concept of solution concentration, as well as gain experience with the chemist's use of this idea.

$$\frac{5}{100} = \frac{50\,000}{1\,000\,000}$$

$$5\% \ (5 \text{ pph}) = 50\,000 \text{ ppm}$$

DESCRIBING SOLUTION CONCENTRATIONS

Building Skills 6

A common intravenous (abbreviated as IV) saline solution used in medicine contains 4.55 g NaCl dissolved in 495.45 g sterilized distilled water. Because a solution is a homogeneous mixture, the NaCl is distributed uniformly throughout the solution. What is the concentration of this solution, expressed as grams NaCl per 100 g solution?

The answer can be calculated in the following way:

$$\frac{4.55 \text{ g NaCl}}{4.55 \text{ g NaCl} + 495.45 \text{ g water}} = \frac{4.55 \text{ g NaCl}}{500 \text{ g solution}}$$

If 500 g solution contains 4.55 g NaCl, then you can determine the answer by calculating how much NaCl is contained in 100 g solution. So 100 g (or 1/5) of the solution will contain 1/5 of the total solute. One-fifth of the total solute is 0.91 g NaCl. Thus 100 g solution contains 0.91 g NaCl and 99.09 g water—1/5 as much as in the full 500-g solution:

$$4.55 \text{ g NaCl} \times 1/5 = 0.91 \text{ g NaCl}$$

$$\frac{0.91 \text{ g NaCl}}{0.91 \text{ g NaCl} + 99.09 \text{ g water}} = \frac{0.91 \text{ g NaCl}}{100 \text{ g solution}} = 0.91\% \text{ NaCl}$$

The concentration of this solution can be expressed as 0.91 g NaCl per 100 g solution. The solution is 0.91% NaCl by mass.

Now consider this example: One teaspoon of sucrose, which has a mass of 10 g, is dissolved in 240 g water. What is the concentration of the solution, expressed as grams sucrose per 100 g solution? As percent sucrose by mass?

> Sucrose, $C_{12}H_{22}O_{11}$, is ordinary table sugar.

Because the solution contains 10 g sucrose and 240 g water, it has a total mass of 250 g. A 100-g solution would contain 2/5 as much solute, or 4 g sucrose. Thus 100 g solution contains 4 g sucrose and 96 g water, a concentration of 4 g sucrose per 100 g solution. To determine the percent sucrose by mass,

$$\frac{10 \text{ g sucrose}}{250 \text{ g solution}} \times 100\% = \frac{4 \text{ g sucrose}}{100 \text{ g solution}} \times 100\% = 4\% \text{ sucrose by mass}$$

1. One teaspoon of sucrose is dissolved in a cup of water. Identify
   a. the solute.
   b. the solvent.

2. What is the concentration of each of the following solutions expressed as percent sucrose by mass?
   a. 17 g sucrose is dissolved in 183 g water.
   b. 30 g sucrose is dissolved in 300 g water.

3. A saturated solution of potassium chloride is prepared by adding 45 g KCl to 100 g water at 60 °C.
   a. What is the concentration of this solution?
   b. What would be the new concentration if 155 g water were added?

4. How would you prepare a "saturated solution" of potassium nitrate (KNO_3)?

C.3 CONSTRUCTING A SOLUBILITY CURVE

Laboratory Activity

Introduction

You have seen and used solubility curves earlier in this unit (pages 46–47). In this activity, you will collect experimental data to construct a solubility curve for succinic acid ($C_4H_6O_4$), a molecular compound. Before you proceed, think about how your knowledge of solubility can help you gather data to construct a solubility curve.

- How can the properties of a saturated solution be used?
- What temperatures can you investigate?
- How many times should you repeat the procedure to be sure of your results?

Discuss these questions with your partner or laboratory group. Your teacher will then discuss with the class how data will be gathered and will demonstrate safe use of the equipment that will be used.

Safety

Keep the following precautions in mind while performing this laboratory procedure.

- ◆ The succinic acid that you will use is slightly toxic if ingested by mouth, so be sure to wash your hands thoroughly at the end of the laboratory.
- ◆ Never stir a liquid with a thermometer. Always use a stirring rod.
- ◆ Use insulated tongs or gloves to remove a hot beaker from a hot plate. Hot glass burns!
- ◆ Dispose of all wastes as directed by your teacher.

Procedure

1. To make a water bath, add approximately 300 mL water to a 400-mL beaker. Heat the beaker, with stirring, to either 45 °C, 55 °C, or 65 °C, as agreed to in your pre-lab class discussion. Ensure that the student team sharing your hot plate is investigating the same temperature. Carefully remove the beaker (using gloves or beaker tongs) when it reaches the desired temperature. NOTE: Do not allow the water-bath temperature to rise more than five degrees above the temperature that you have chosen. Return the beaker to the hot plate as needed to maintain the appropriate water-bath temperature.

2. Place between 4 g and 5 g succinic acid in each of two test tubes.
△ CAUTION: *Be careful not to spill any of the succinic acid. If you do, clean up and dispose of the succinic acid as directed by your teacher.* Add 20.0 mL distilled water to each test tube.

3. Place each test tube in the water bath and take turns stirring the succinic acid solution with a glass stirring rod every 30 seconds for 7 minutes. Each minute, place the thermometer in the test tube and monitor the temperature of the succinic acid solution, ensuring that it is within 2 °C of the temperature that you have chosen.

4. At the end of 7 minutes, carefully decant the clear liquid from each test tube into a separate, empty test tube, as demonstrated by your teacher.

5. Carefully pour the hot water from the beaker into the sink and fill the beaker with water and ice.

6. Place the two test tubes containing the clear liquid in the ice bath for 2 minutes. Stir the liquid in each test tube gently once or twice. Remove the test tubes from the ice water. Allow the test tubes to sit at room temperature for 5 minutes. Observe each test tube carefully during that time. Record your observations.

7. Tap the side of each test tube and swirl the liquid once or twice to cause the crystals to settle evenly on the bottom of the test tubes.

8. Measure the height of crystals collected (in millimeters, mm). Have your partner(s) measure the crystal sample height and compare your results. Report the average crystal height for your two test tubes to your teacher.

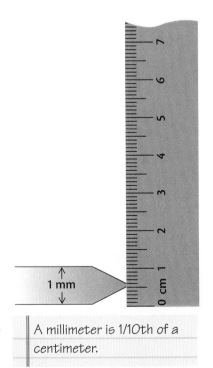

A millimeter is 1/10th of a centimeter.

9. Rinse the succinic acid crystals from the test tubes into a collection beaker designated by your teacher. Make sure that your laboratory area is clean.

10. Wash your hands thoroughly before leaving the laboratory.

Data Analysis

1. Find the mean crystal height obtained by your class for each temperature reported.

2. Plot the mean crystal height in millimeters (*y* axis) versus the water temperature in degrees Celsius (*x* axis).

Questions

1. Why is it important to collect data from more than one trial at a particular temperature?

2. How did you make use of the properties of a saturated solution at different temperatures?

3. Did all the succinic acid that originally dissolved in the water crystallize out of the solution? Explain your answer.

4. Given the pooled class data, did you have enough points to make a reliable solubility curve for succinic acid? Would the curve be good enough to make useful predictions about succinic acid solubility at temperatures not investigated in this activity? Explain your answer.

5. What procedures in this activity could lead to errors? How would each error affect your data?

6. Using your knowledge of solubility, propose a different procedure for gathering data to construct a solubility curve.

C.4 DISSOLVING IONIC COMPOUNDS

You have just investigated the process of a compound dissolving in water. What you observed is called a macroscopic phenomenon. However, what chemistry is primarily concerned with is what happens at the submicroscopic level—atomic and molecular phenomena that are not easily observed. As you have seen, temperature, agitation, and time all contribute to dissolving a solid material. But how do the atoms and molecules of the solute and solvent interact to make this happen?

Experiments suggest that water molecules are electrically **polar.** Although the entire water molecule is electrically neutral, the electrons are not evenly distributed in its structure. A polar molecule has an uneven distribution of electrical charge, which means that each molecule has a positive region on one end and a negative region on the other end. Evidence also suggests that a water molecule has a bent or V-shape, as illustrated in Figure 21 on page 52, rather than a linear, sticklike shape as in H–O–H. The "oxygen end" is an electrically negative region that has a greater

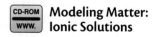
Modeling Matter: Ionic Solutions

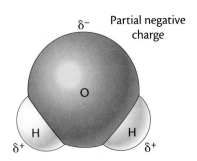

δ−   Partial negative charge

O

δ+   H   H   δ+

Partial positive charge

Figure 21 *Polarity of a water molecule. The δ+ and δ− indicate partial electrical charges.*

The water solubility of some ionic compounds can be extremely low indeed. For example, at room temperature, lead(II) sulfide, PbS, has a solubility of only about 10^{-14} g (0.00000000000001 g) per liter of water solution.

concentration of electrons (shown as δ−) compared with the two "hydrogen ends," which are electrically positive (shown as δ+). The Greek symbol δ (delta) means "partial"—thus partial plus and partial minus electrical charges are indicated. Because these charges balance, the molecule as a whole is electrically neutral.

Polar water molecules are attracted to other polar substances and to substances composed of electrically charged particles. These attractions make it possible for water to dissolve a great variety of substances.

One way to imagine the process of dissolving a substance in water is to liken it to a tug of war. Many solid substances, especially ionic compounds, are crystalline. In ionic crystals, positively charged cations are surrounded by negatively charged anions, with the anions likewise surrounded by cations. The crystal is held together by attractive forces between the cations and the anions. The substance will dissolve only if its ions are so strongly attracted to water molecules that the water "tugs" the ions from the crystal.

Water molecules are attracted to ions located on the surface of an ionic solid, as shown by the models in Figure 22a. The water molecule's negative (oxygen) end is attracted to the crystal's positive ions. The positive (hydrogen) ends of other water molecules are attracted to the negative ions of the crystal. When the attractive forces between the water molecules and the surface ions are strong enough, the bonds between the crystal and its surface ions become strained, and the ions may be pulled away from the crystal. Figure 22b uses models of water molecules and solute ions to illustrate the results of water "tugging" on solute ions. The detached ions become surrounded by water molecules, producing a water solution, as shown in Figure 22c.

Using the description and illustrations of this process, can you determine what influences whether an ionic solid will dissolve? Because dissolving entails competition among three types of attractions—those between solvent and solute particles, between solvent particles themselves, and between particles within the solute crystals—the properties of both solute and solvent affect whether two substances will form a solution. Water is highly polar, so it will be effective at dissolving charged or ionic substances. However, if positive–negative attractions between cations and anions in the crystal are sufficiently strong, a particular ionic compound may be only slightly soluble in water.

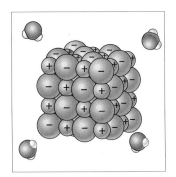

Figure 22a *Polar water molecules are attracted to the ions in an ionic crystal.*

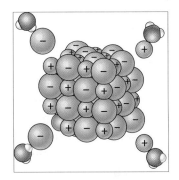

Figure 22b *Ions from the crystalline solid are pulled away by water molecules.*

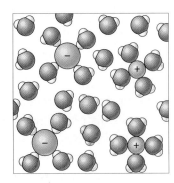

Figure 22c *A solution is formed when detached ions become surrounded by water molecules.*

MODELING MATTER

DISSOLVING IONIC COMPOUNDS

You have now learned about solubility, solubility curves, and the process of dissolving ionic compounds in water. As part of these discussions, visual models, such as those presented in Figure 22 (page 52) have been used to describe the process of dissolving. In this activity, you will combine these models with your knowledge of solubility curves to create new models of ions dissolved in water.

1. Suppose you dissolved 40 g potassium chloride (KCl) in 100 g water at 50 °C. You then let the solution cool to room temperature, about 25 °C.

    a. What changes would you see in the beaker as the solution cooled? See Figure 23.

    b. Draw models of what the contents in the beaker would look like at the molecular level at 50 °C, 40 °C, and 25 °C. Keep these points in mind:

    ◆ You will need to consider whether the sample at each temperature is saturated, unsaturated, or supersaturated and draw the model accordingly. The solubility curve in Figure 23 will be helpful.

    ◆ It is impossible to draw all the ions and molecules in this sample. The ions and molecules that you draw will represent what is happening on a much larger scale.

2. An unsaturated solution will become more concentrated if you add more solute. Decreasing the total volume of water in the solution (such as by evaporation) also causes an increase in the solution's concentration. Consider a solution made by dissolving 20 g KCl in 100 g water at 40 °C.

    a. Draw a model of this solution.

    b. Suppose that while the solution was kept at 40 °C, 25% of the water evaporated.
    i. Draw a model of this solution and describe how it differs from the model of the original solution.
    ii. How much water must evaporate at this temperature to cause the first potassium chloride crystals to form?

3. A solution may be diluted (made less concentrated) by adding water.

    a. Draw a model of a solution containing 10 g KCl in 100 g water at 25 °C.

    b. Suppose you diluted this solution by adding another 100 g water with stirring. Draw a model of this new solution.

    c. Compare your drawings in 3a and 3b. What key feature is different in the two models? Why?

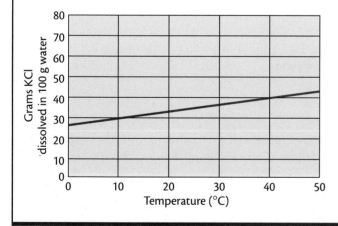

Figure 23 *Solubility curve for potassium chloride.*

Now that you know how solutions of ionic compounds are formed and described, it is time to consider some possible causes of the Riverwood fish kill. The information that you are about to read suggests some possible "culprits"—substances that can dissolve or be suspended in water and harm living things. These substances include heavy metals, acids and bases, molecular substances, and dissolved oxygen gas. Although all of these substances are normally found in natural water sources, the levels at which they are present can positively or negatively affect aquatic life.

C.5 INAPPROPRIATE HEAVY-METAL ION CONCENTRATIONS IN RIVER?

Many metal ions, such as iron(II) (Fe^{2+}), potassium (K^+), calcium (Ca^{2+}), and magnesium (Mg^{2+}), are essential to the health of humans and other organisms. For humans, these ions are obtained primarily from foods, but they may also be present in drinking water.

Not all metal ions that dissolve in water are beneficial, however. Some heavy-metal ions, called heavy metals because their atoms have greater mass than the masses of essential metallic elements, are harmful to humans and other organisms. Among the heavy-metal ions of greatest concern in water are cations of lead (Pb^{2+}) and mercury (Hg^{2+}. Lead and mercury are particularly likely to cause harm because they are widely used and dispersed in the environment. Heavy-metal ions are toxic because they bind to proteins in biological systems (such as your body), preventing the proteins from performing their intended tasks. As you might expect, because proteins play many important roles in body functioning, heavy-metal poisoning effects are severe. They include damage to the nervous system, brain, kidneys, and liver and even death.

Unfortunately, heavy-metal ions are not removed as waste as they move up through the food chain. They become concentrated within the bodies of fish and shellfish, even when their abundance in the surrounding water is only a few parts per million. Such aquatic creatures then become hazardous for humans and other animals to consume.

In very low concentrations, heavy-metal ions are hard to detect in water and even more difficult and costly to remove. So how can heavy-metal poisoning be prevented? One of the easiest and most effective ways is to prevent the heavy-metal ions from entering water systems in the first place. This prevention can be accomplished by producing and using alternate materials that do not contain these ions and thus are not harmful to health or the environment. Such practices, which prevent pollution by eliminating the production and/or use of hazardous substances, are classified as examples of **green chemistry.** Such practices are applicable to heavy metals and to many other types of pollution.

> The concentration of substances as they move through the food chain is known as bioaccumulation.

Lead (Pb)

> Locate lead on the Periodic Table at the back of your textbook. Compare its location with those of the essential metal ions that you just read about.

Lead is probably the heavy metal most familiar to you. Its symbol, Pb, is based on the element's original Latin name *plumbum,* also the source of the word "plumber."

Lead and lead compounds have been, and in some cases still are, used in pottery, automobile electrical storage batteries, solder, cooking vessels, pesticides, and paints. One compound of lead and oxygen, red lead (Pb_3O_4), is the primary ingredient in paint that protects bridges and other steel structures from corrosion.

Although lead water pipes were used in the United States in the early 1800s, they were replaced by iron pipes after it was discovered that water transported through lead pipes could cause lead poisoning. Romans constructed lead water pipes more than 2000 years ago; some of them are still in working condition. In modern homes, copper or plastic water pipes are used to prevent any contact between household water and lead.

Until the 1970s, the molecular compound tetraethyl lead, $Pb(C_2H_5)_4$, was added to gasoline to produce a better-burning automobile fuel. Unfortunately, the lead entered the atmosphere through automobile exhaust as lead oxide. Although the phaseout of leaded gasoline has reduced lead emissions, lead contamination remains in the soil surrounding heavily traveled roads. In some homes built before 1978 and not since repainted, the flaking of old leaded paint is another source of lead poisoning, particularly among children who may ingest the flaking paint.

Mercury (Hg)

Mercury is the only metallic element that is a liquid at room temperature. In fact, its symbol comes from the Latin *hydrargyrum*, meaning quick silver or liquid silver.

Mercury has several important uses, some due specifically to its liquid state. It is an excellent electrical conductor, so it is used in "silent" light switches. It is also found in medical and weather thermometers, thermostats, mercury-vapor street lamps, fluorescent light bulbs, and some paints. Elemental mercury can be absorbed directly through the skin, and its vapor is quite hazardous to health. At room temperature, there will always be some mercury vapor present if liquid mercury is exposed to air, so any direct exposure to mercury is best avoided.

Because mercury compounds are toxic, they are useful in eliminating bacteria, fungi, and agricultural pests when used in antiseptics, fungicides, and pesticides. In the eighteenth and nineteenth centuries, mercury(II) nitrate, $Hg(NO_3)_2$, was used in making the felt hats popular at that time. After unintentionally absorbing this compound through their skin for several years, hat makers often suffered from mercury poisoning. Their symptoms included numbness, staggered walk, tunnel vision, and brain damage, thus giving rise to the expression "mad as a hatter."

The sudden release of a large amount of heavy-metal ions might cause a fish kill—depending on the particular metal ion, its concentration, the species of fish present, and other factors. Was such a release responsible for the Riverwood fish kill? As you read about other possible causes of the fish kill, keep in mind some questions that are relevant to all of the potential culprits. Is there a source of this substance along the Snake River near the site of the fish kill? What concentration of this solute would be toxic to various species of fish?

> Locate mercury on the Periodic Table. Make a prediction about the locations of heavy metals on the Periodic Table.

C.6 INAPPROPRIATE pH LEVELS IN RIVER?

You have likely heard the term pH used before, perhaps in connection with acid rain or hair shampoo. What is pH, and could it possibly help account for the fish kill in Riverwood? The **pH scale** is a convenient way to measure and report the acidic, basic, or chemically neutral character of a solution.

Nearly all pH values are in the range from 0 to 14, although some extremely acidic or basic solutions may be outside this range. At room temperature, any pH values less than 7 indicate an acidic condition; the lower the pH, the more acidic the solution. Solutions with pH values greater than 7 are basic; the higher the pH, the more basic the solution. Basic solutions are also called alkaline solutions. Quantitatively, a change of one pH unit indicates a tenfold difference in acidity or alkalinity. For example, lemon juice, with a

Some Common Acids and Bases		
Name	Formula	Use
Acids		
Acetic acid	$HC_2H_3O_2$	In vinegar (typically a 5% solution of acetic acid)
Carbonic acid	H_2CO_3	In carbonated soft drinks
Hydrochloric acid	HCl	Used in removing scale buildup from boilers and for cleaning materials
Nitric acid	HNO_3	Used in the manufacture of fertilizers, dyes, and explosives
Phosphoric acid	H_3PO_4	Added to some soft drinks to give a tart flavor; also used in the manufacture of fertilizers and detergents
Sulfuric acid	H_2SO_4	Largest-volume substance produced by chemical industry; present in automobile battery fluid
Bases		
Calcium hydroxide	$Ca(OH)_2$	Present in mortar, plaster, and cement; used in paper pulping and dehairing animal hides
Magnesium hydroxide	$Mg(OH)_2$	Active ingredient in milk of magnesia
Potassium hydroxide	KOH	Used in the manufacture of some liquid soaps
Sodium hydroxide	NaOH	A major industrial product; active ingredient in some drain and oven cleaners; used to convert animal fats into soap

Figure 24 *The name, formula, and common use of some familiar acids and bases.*

pH of about 2, is nearly ten times as acidic as soft drinks, which have a pH of about 3.

Acids and bases, some examples of which are listed in Figure 24, can also be identified by certain chemical properties. For example, the vegetable dye litmus turns blue in a basic solution and red in an acidic solution. Both acidic and basic solutions conduct electricity. Each type of solution has a distinctive taste and a distinctive feel on your skin. (**CAUTION:** *You should never test these sensory properties in the laboratory.*) In addition, concentrated acids and bases are able to react chemically with many other substances. You are probably familiar with the ability of acids and bases to corrode, or wear away, other materials. Corrosion is a type of chemical reaction.

Most acid molecules have one or more hydrogen atoms that can be released rather easily in water solution. These "acidic" hydrogen atoms are usually written first in the formula for an acid. See Figure 24.

Many bases are ionic substances that include hydroxide ions (OH^-) in their structures. Sodium hydroxide, NaOH, and barium hydroxide, $Ba(OH)_2$, are two examples. Some bases, such as ammonia (NH_3) and baking soda (sodium bicarbonate, $NaHCO_3$), contain no OH^- ions but still produce basic solutions because they react with water to generate OH^- ions, as illustrated by the following equation.

$$NH_3 \quad + \quad H_2O \quad \rightarrow \quad NH_4^+ \quad + \quad OH^-$$

What about substances that display neither acidic nor basic characteristics? Chemists classify these substances as chemically neutral. Water, sodium chloride (NaCl), and table sugar (sucrose, $C_{12}H_{22}O_{11}$) are all examples of chemically neutral compounds.

At 25 °C, a pH of 7 indicates a chemically neutral solution. The pH values of some common materials are shown in Figure 25.

As you can see in Figure 25, rainwater is naturally slightly acidic. This is because the atmosphere contains certain substances—carbon dioxide (CO_2) for one—that produce acidic solutions when dissolved in water. Both acidic and basic solutions have effects on living organisms—effects that depend on the pH of the water. When the pH of water is too low (meaning high acidity), fish-egg development is impaired, thus hampering the ability of fish to reproduce. Water solutions with low (acidic) pH values also tend to increase the concentrations of metal ions in natural waters by leaching the metals from surrounding soil. These metal ions can include aluminum ions (Al^{3+}), which are toxic to fish when present in sufficiently high concentration. High pH (basic contamination) is a problem for living organisms primarily because alkaline solutions are able to dissolve organic materials, including skin and scales.

The U.S. Environmental Protection Agency (EPA) requires drinking water to be within the pH range from 6.5 to 8.5. However, most fish can

Vinegar is an acid that you have tasted; that common kitchen ingredient is considered a dilute solution of acetic acid.

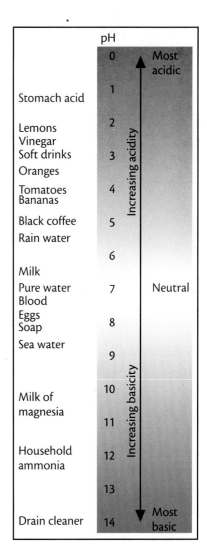

Figure 25 *The pH values of some common materials.*

tolerate a slightly wider pH range, from about 5 to 9, in lake or river water. Serious freshwater anglers try to catch fish in water between pH 6.5 and 8.2.

On a normal day, the pH of the water in the Snake River in Riverwood ranges between 7 and 8, nearly optimal for freshwater fishing. Could the pH have changed abruptly, killing the fish? Was acidic or basic contamination responsible for the Riverwood crisis?

C.7 INAPPROPRIATE MOLECULAR-SUBSTANCE CONCENTRATIONS IN RIVER?

Solubility of Molecular Substances [CD-ROM / WWW.]

Until now, the types of substances considered suspects in the Riverwood mystery have been ionic substances, those that dissolve in water to release ions. Are there other types of substances that dissolve in water and possibly present a hazard to aquatic life? Some substances, such as sugar and ethanol, dissolve in water but not in the form of ions. These substances belong to a class of materials known as **molecular substances** because they are composed of molecules.

Unlike ionic substances, which are crystalline solids at normal conditions, molecular substances can be found as solids, liquids, or gases at room temperature. Some molecular substances such as oxygen (O_2) and carbon dioxide (CO_2) have little attraction between their molecules and are thus gases at normal conditions. Molecular substances such as ethanol (ethyl alcohol, C_2H_5OH) and water (H_2O) have larger between-molecule attractions, causing these "stickier" molecules to form liquids at normal conditions. Other molecular substances with even greater between-molecule attractions—succinic acid ($C_4H_6O_4$), for example—are solids at normal conditions. Their stronger attractive forces hold the molecules together more tightly, in effect determining in which state the substance will be found.

You investigated the solubility behavior of succinic acid earlier. See p. 49.

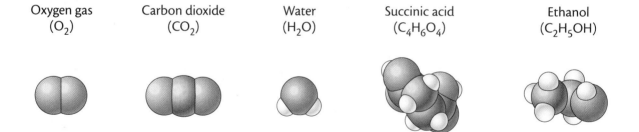

Oxygen gas (O_2)   Carbon dioxide (CO_2)   Water (H_2O)   Succinic acid ($C_4H_6O_4$)   Ethanol (C_2H_5OH)

What determines how soluble a molecular substance will be in water? The attraction of a substance's molecules for each other compared with their attraction for water molecules plays a major part. But what causes these attractions? The distribution of electrical charge within molecules has a great deal to do with it.

Most molecular compounds contain atoms of nonmetallic elements. As you learned earlier (page 32), these atoms are linked together by the attraction of one atom's positively charged nucleus for another atom's negatively charged electrons. If differences in electron attraction between atoms are large enough, electrons move from one atom to another, forming ions. This is what often happens between a metallic atom and a nonmetallic atom when an ionic compound is formed. An atom's ability to attract shared electrons in its bonding within a substance is known as its **electronegativity**. In molecular substances, these differences in electron attraction, or electronegativities, are not large enough to cause ions to form, but they may cause the electrons to be unevenly distributed among the atoms.

You already know that the "oxygen end" of a water molecule is electrically negative compared with its positive "hydrogen end." That is, water molecules are polar and serve as the most common example of a polar solvent. Such charge separation (and resulting molecular polarity) is found in many molecules whose atoms have sufficiently different electronegativities.

Polar molecules tend to dissolve readily in polar solvents such as water. For example, water is a good solvent for sugar and ethanol, both composed of polar molecules. Similarly, nonpolar liquids are good solvents for other nonpolar molecules. Nonpolar cleaning fluids are used to "dry clean" clothes because they readily dissolve nonpolar body oils found in fabric. In contrast, nonpolar molecules (such as those of oil and gasoline) do not dissolve well in polar solvents (such as water or ethanol).

This pattern of solubility behavior—polar substances dissolving in polar solvents, nonpolar substances dissolving in nonpolar solvents—is often summarized in the generalization "Like dissolves like." This rule also explains why nonpolar liquids are usually ineffective in dissolving ionic and polar molecular substances.

Were dissolved molecular substances present in the Snake River water where the fish died? Most likely yes; at least in small amounts. Were they responsible for the fish kill? That depends on which molecular substances were present and at what concentrations. And that, in turn, depends on how each solute interacts with water's polar molecules.

In the following laboratory activity, you will investigate and compare the solubility behavior of some typical molecular and ionic substances.

> Unfortunately, many nonpolar dry-cleaning solvents are damaging to both human health and the environment. However, the recent development of new technologies allows environmentally benign nonpolar solvents such as liquefied carbon dioxide to be used in the dry-cleaning industry.

> Various molecular substances may normally be present at very low levels—so low that no harm to living things is observed.

C.8 SOLVENTS

Introduction

The *Riverwood News* reported earlier that Dr. Brooke believes that a substance dissolved in the Snake River is one likely fish-kill cause. She based her judgment on her chemical knowledge and experiences with water and other substances. Dr. Brooke also has a general idea about which contaminating solutes she can initially rule out: those that cannot dissolve appreciably in water. Such background knowledge helps Dr. Brooke (and other chemists) reduce the number of water tests required in the laboratory.

What, then, do the terms "soluble" and "insoluble" mean? Is anything truly insoluble in water? It is likely that at least a few molecules or ions of any substance will dissolve in water. Thus the term "insoluble" actually refers to substances that are only very, very slightly soluble in water. Chalk, for example, is considered insoluble in water, even though 1.53 mg calcium carbonate ($CaCO_3$, the main ingredient in chalk) can dissolve in 100 g water at 25 °C.

In this laboratory activity, you will first investigate the solubilities of various molecular and ionic solutes in water. These solubility data, along with toxicity data, will help you rule out some solutes as likely causes of the fish kill. You will then test other solvents and examine the solubility data for any general patterns.

Your teacher will tell you which particular solutes you will investigate. List them in your laboratory notebook.

Part I: Designing a Procedure for Investigating Solubility in Water

Your teacher will direct you to discuss with either the whole class or your laboratory partner a procedure for testing the room-temperature solubilities of the substances listed in your laboratory notebook. (If you have performed solubility tests before, it may be useful to recall how you did them.) With your partner, design a step-by-step investigation that will allow you to determine whether each solute is soluble (S), slightly soluble (SS), or insoluble (I) in room-temperature water.

The following questions will help you design your procedure.

1. What particular observations will allow you to judge how well each solute dissolves in the polar solvent water? That is, how will you decide whether to classify a given solute as soluble, slightly soluble, or insoluble in water?

2. Which variables will need to be controlled? Why?

3. How should the solute and solvent be mixed—all at once or a little at a time? Why?

In designing your procedure, keep these concerns in mind.

◆ Avoid any direct contact of your skin with any solutes.

◆ Follow your teacher's directions for waste disposal.

When you and your partner have agreed on a written procedure, get it approved by your teacher. Construct a data table for your results and you are ready for Part II.

Part II: Investigating Solubility in Water

Use your approved procedure to investigate the solubility in water of the listed substances. Record the data in your data table.

Part III: Investigating Solubility in Ethanol and Lamp Oil

It is clear that the task of determining what may have caused the fish kill can be simplified somewhat by focusing efforts on substances that will dissolve appreciably in water. However, in dealing with other solubility-based problems, chemists sometimes find it helpful to use solvents other than water—ethanol and lamp oil serve that role in this activity.

You and your partner will investigate the solubility of some or all of the solutes from Part II in ethanol and lamp oil. You should also test the solubility of water in ethanol and in lamp oil. By gaining experiences with three liquid solvents—lamp oil, ethanol, and water—you will be prepared to recognize some general patterns regarding solubility behavior.

Can you use the same procedure that you designed for Part II? If not, what parts of that procedure should be revised? In considering your Part III procedure, again keep the questions listed on page 60 in mind.

Have your proposed procedure approved by your teacher. Before you start the laboratory work, test your interpretation of the results of Part II by predicting what you think you will observe regarding solubility in each case. Include these predictions in your data table for Part III. Then collect and record your data for both solvents.

Wash your hands thoroughly before leaving the laboratory.

Questions

Part II

1. According to your data, which tested solutes are least likely to be dissolved in the Snake River? Why?

2. Compare your data with those of the rest of the class. Are there any differences? If so, how can those differences be explained?

Part III

3. a. How does the behavior of ethanol as a solvent compare with that of water?
   b. How does ethanol's behavior compare with that of lamp oil?

4. a. Were any of your solubility observations unexpected?
   b. If so, explain what you expected, why it was expected, and how your expectations compare with what you actually observed.

5. Based on your data, what general pattern of solubility behavior can you summarize and describe?

6. Predict the solubility behavior of each solid solute in:
   a. hexane, a liquid that is essentially insoluble in water.
   b. ethylene glycol, a liquid that is very soluble in ethanol.

7. a. Given that water is a polar solvent and lamp oil is a nonpolar solvent, classify each molecular solute tested as polar or nonpolar.
   b. How did you decide?

8. How useful is the rule "like dissolves like" for predicting solubility? Explain your answer on the basis of your results.

9. In Part II, water was the "solvent," but in Part III, water was a "solute."
   a. How can it be both?
   b. How do you know whether a substance is a solute or a solvent?

> The familiar saying "oil and water do not mix" has a chemical basis!

You now know about the solubility of some solids and liquids in water. As you read on to learn about the behavior of gases in natural waters, consider the possibility that a dissolved gas was responsible for the Snake River catastrophe.

C.9 INAPPROPRIATE DISSOLVED OXYGEN LEVELS IN RIVER?

You have noted that the solubility of ionic and molecular solids in water usually increases when the water temperature is raised. Do gases behave similarly to solids in solution? Look at Figure 26, which shows the solubility curve for oxygen gas, plotted as milligrams oxygen dissolved per 1000 g water.

What is the solubility of oxygen gas in 20 °C water? In 40 °C water? As you can see, increasing the water temperature causes the gas to be less soluble! Note also the magnitude of the values for oxygen solubility. Compare these values with those for solid solutes as shown in Figure 20. At 20 °C, about 30 g potassium nitrate will dissolve in 100 g water. In contrast, only about 9 mg (0.009 g) oxygen gas will dissolve in ten times as much water—1000 g water—at this temperature. It should be clear that most gases are far less soluble in water than are many ionic solids.

When you considered the solubility of molecular and ionic solids in water, you found that solubility depended on two factors—temperature and the nature of the solvent. The solubility of a gas depends on these two factors as well. But it also depends on gas pressure. Referring to Figure 27, note what happens to oxygen's solubility as the pressure of oxygen gas above it is increased. Does more or less gas dissolve in the same amount of water? What happens if the gas pressure is doubled? For example, look at the solubility of O_2 at one atmosphere and at two atmospheres of pressure. Also consider the shape of the graph line. What type of relationship does a linear graph line indicate?

As you have by now deduced, gas solubility in water is directly proportional to the pressure of that gaseous substance on the liquid. You see one effect of this proportionality every time you open a can or bottle of carbonated soft drink. As the gas pressure on the liquid is reduced by opening the container to the air, some dissolved carbon dioxide gas (CO_2) escapes from the liquid in a rush of bubbles.

Because there is not very much $CO_2(g)$ present in air, carbon dioxide gas must be forced into the carbonated beverage at high pressure just before

> The metric unit for pressure is the pascal (Pa): 1 atmosphere (atm) = 101 325 Pa

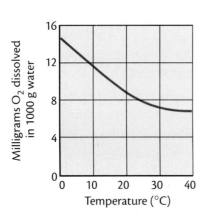

Figure 26 Solubility curve for O_2 gas in water in contact with air.

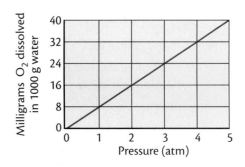

Figure 27 Relationship between solubility and pressure of O_2 at 25 °C.

the container is sealed. This increases the amount of carbon dioxide that can dissolve in the beverage. When the can or bottle is opened, CO_2 gas pressure on the liquid drops back to its usual low level. Dissolved carbon dioxide gas escapes from the liquid until it reaches its (lower) solubility at this lower pressure. When the fizzing stops, you might describe the beverage as having "gone flat." Actually, the excess carbon dioxide gas has simply escaped into the air; the resulting solution is still saturated with CO_2 at the new conditions.

C.10 TEMPERATURE, DISSOLVED OXYGEN, AND LIFE

On the basis of what you know about the effect of temperature on solubility, you may be wondering if the temperature of the Snake River had something to do with the fish kill. As you just learned, water temperature affects how much oxygen gas can dissolve in the water. Various forms of aquatic life, including the many species of fish, have different requirements for the concentration of dissolved oxygen needed to survive. Figure 28 contains this information. How, then, does a change in the temperature of the natural waters affect the fish internally?

Fish are "cold-blooded" animals; their body temperatures rise and fall with the surrounding water temperature. If the water temperature rises, the body temperatures of fish also rise. This increase in body temperature in turn increases fish metabolism, a complex series of interrelated chemical reactions that keep fish alive. As these internal reaction rates speed up, the fish eat more, swim more, and require more dissolved oxygen. The rate of metabolism also increases for other aquatic organisms, such as aerobic bacteria, that compete with fish for dissolved oxygen.

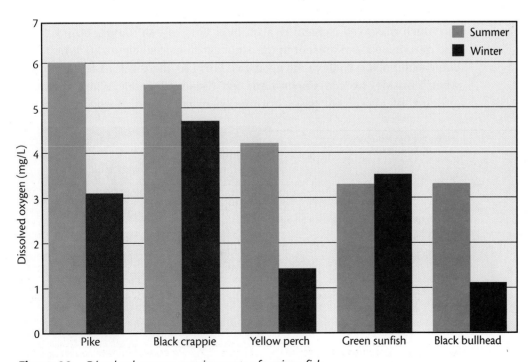

Figure 28 *Dissolved-oxygen requirements of various fish.*

Figure 29 *Maximum water temperature tolerance in fish.*

Maximum Water Temperature Tolerance in Fish (24-hour Exposure)		
	Maximum Temperature	
Fish	°C	°F
Trout (brook, brown, rainbow)	24	75
Channel catfish	35	95
Lake herring (cisco)	25	77
Largemouth bass	34	93
Northern pike	30	86

As you can now see, an increase in water temperature affects fish both externally and internally. A long stretch of hot summer days sometimes results in large fish kills, in which hundreds of fish literally suffocate. Figure 29 summarizes the maximum water temperatures at which selected fish species can survive.

Sometimes hot summer days are not to blame. Often, high lake or river water temperatures can be traced to human activity. Many industries, such as electrical power generation, depend on natural bodies of water to cool heat-producing processes. Cool water is drawn from lakes or rivers into an industrial or power-generating plant, and devices called heat exchangers transfer thermal energy (heat energy) from the processing area to the cooling water. The heated water is then released back into the lakes or rivers, either immediately or after the water has partly cooled. If the water is too warm, it can upset the balance of life in lakes and rivers.

At this point, it should be clear that there is a lower limit on the amount of dissolved oxygen needed for fish to survive. Is there an upper limit? Can too much dissolved oxygen be a problem for fish? In nature, both oxygen and nitrogen gas are present in the air at all times. See Figure 30. When oxygen gas dissolves, so does nitrogen gas. This fact turns out to be significant when considering the upper limit for dissolved gases. When the total amount of dissolved gases—primarily oxygen and nitrogen—reaches

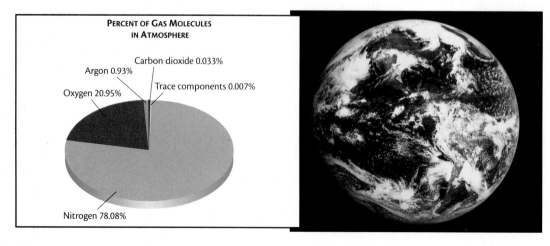

Figure 30 *The percent composition of the atmosphere—the envelope of gases that surrounds planet Earth.*

between 110% and 124% of saturation (a state of supersaturation), a condition called gas-bubble trauma may develop in fish.

This situation is dangerous because the supersaturated solution causes gas bubbles to form in the blood and tissues of fish. Oxygen-gas bubbles can be partially utilized by fish during metabolism, but nitrogen-gas bubbles block the flow of blood within the fish. This blockage results in the death of the fish within hours or days. Gas-bubble trauma can be diagnosed by noting gas bubbles in the gills of dead fish if they are dissected promptly after death. Supersaturation of water with nitrogen and oxygen gases can occur at the base of a dam or a hydroelectric project as released water forms "froth," trapping large quantities of air.

So, back to the original question: What caused the Riverwood fish kill? You have considered several possible causes. How will you determine which one was the actual cause? You will start by examining water-related data collected by scientists and engineers on the Snake River. From these data and what you have learned, you will make your own decision.

C.11 DETERMINING THE CAUSE OF THE FISH KILL

Making Decisions

Snake River watershed data have been collected and monitored since the early 1900s. Although some measurements and methods have changed during that time, excellent data have been gathered, particularly in recent years. Data are available regarding the following factors.

- water temperature and dissolved oxygen
- rainfall
- water flow
- dissolved molecular substances
- heavy metals
- pH
- nitrate and phosphate levels
- organic carbons

Your teacher will assign you to a group to study some of the data just listed. Each group will complete data-analysis procedures for its assigned data. After groups have finished their work, the class will share their analyses and draw conclusions about factors possibly related to the fish kill.

The following background information will help you complete the analysis of the data assigned to you by your teacher.

DATA ANALYSIS

Introduction

Interpreting graphs of environmental data requires a slightly different approach from the one you took in interpreting a solubility curve (page 47). Rather than seeking a predictable relationship, you will be looking for regularities or patterns among the values. Any major irregularity in the data may suggest a problem related to the water factor being evaluated. The following suggestions will help you prepare and interpret such graphs.

- Choose your scale so that the graph becomes large enough to fill most of the available space on the graph paper.

- Assign each regularly spaced division on the graph paper some convenient, constant value. The graph-paper line interval should have a value easily "divided by eye," such as 1, 2, 5, or 10, rather than a value such as 6, 7, 9, or 14.

- An axis scale does not have to start at "zero," particularly if the plotted values cluster in a narrow range not near zero.

- Label each axis with the quantity and unit being graphed.

- Plot each point. Then draw a small symbol around each point, like this: ⊙. If you plot more than one set of data on the same graph, distinguish each set of points by using a different color or geometric shape, such as: ▱, ▽, or △.

- Give your graph a title that will readily convey its meaning and purpose to a reader.

- If you use technology—such as a graphing calculator or computer program—to prepare your graphs, ensure that you follow the guidelines just given. Different devices or programs have different ways of processing data. Choose the appropriate type of graph (scatter plot, bar graph, and so forth) for your data.

Procedure

1. Prepare a graph for each of your group's Snake River data sets. Label the *x*, or independent, axis with the consecutive months indicated in the data. Label the *y*, or dependent, axis with the water factor measured, accompanied by its units.

2. Plot each data point and connect the consecutive points with straight lines.

Questions

1. Is any pattern apparent in your group's plotted data?

2. Can you offer possible explanations for any pattern or irregularities that you detect?

3. Do you think the data analyzed by your group might help to account for the Snake River fish kill? Why? How?

4. Prepare to share your group's findings regarding Questions 1 through 3 in a class discussion. During the class discussion, take notes on key findings reported by each data-analysis group. Also note and record significant points raised in the data-analysis discussions.

Questions & Answers CD-ROM WWW.

Your class will reassemble several times during your study of Section D to discuss and consider implications of the water-analysis data that you have just processed. In particular, guided by the patterns and irregularities found in your analysis of Snake River data, your class will seek an explanation or scenario that accounts for the observed data and for the resulting Riverwood fish kill. Good luck!

SECTION SUMMARY

Reviewing the Concepts

♦ **The solubility of a substance in water can be expressed as the quantity of that substance that will dissolve in a certain quantity of water at a specified temperature.**

1. If the solubility of sugar (sucrose) in water is 2.0 g/mL at room temperature, what is the maximum amount of sugar that will dissolve in 946 mL (1 quart) water?

2. Explain why three teaspoons of sugar will completely dissolve in a serving of hot tea, but not in an equally sized serving of iced tea.

3. Rank the substances in Figure 20 (page 46) from most soluble to least soluble

   a. at 20 °C.

   b. at 80 °C.

♦ **Solutions can be described qualitatively or quantitatively. In qualitative terms, solutions can be classified as unsaturated, saturated, or supersaturated. In quantitative terms, the concentration of a solution expresses the relative quantities of solute and solvent in a particular solution.**

4. A 35-g sample of ethanol is dissolved in 115 g water. What is the concentration of the ethanol, expressed as grams ethanol per 100 g solution?

5. Calculate the masses of water and sugar in a 55-g sugar solution that is labeled 20.0% sugar.

6. Using the graph on page 46, answer the following questions about the solubility of potassium nitrate, KNO_3.

   a. What is the maximum mass of KNO_3 that can dissolve in 100 g water if the water temperature is 20 °C?

   b. At 30 °C, 55 g KNO_3 is dissolved in 100 g water. Is this solution saturated, unsaturated, or supersaturated?

   c. A saturated solution of KNO_3 is formed in 100 g water at 75 °C. If the saturated solution is cooled to 40 °C, how many grams of solid KNO_3 should form?

♦ **Polar bonds have an uneven distribution of electrical charge. Polar O–H bonds in water help explain water's ability to dissolve many ionic solids.**

7. Draw a model that shows how molecules in liquid water generally arrange themselves relative to one another.

8. Why does table salt (NaCl) dissolve in water but not in cooking oil?

♦ **Heavy-metal ions, such as Pb^{2+} and Hg^{2+}, are useful resources but can be toxic if introduced into biological systems, even in small amounts.**

9. a. What are heavy metals?

   b. List some general effects of heavy-metal poisoning.

10. What are some possible sources of human exposure to heavy metals?

11. What items might be found in an urban landfill that could contribute heavy-metal ions to groundwater?

• Water solutions can be characterized as acidic, basic, or chemically neutral on the basis of their chemical properties.

12. Classify each sample as acidic, basic, or chemically neutral:

   a. seawater (pH = 8.6)
   b. drain cleaner (pH = 13.0)
   c. vinegar (pH = 2.7)
   d. pure water (pH = 7.0)

13. Which is more acidic, a tomato or a soft drink? (See Figure 25 on page 57)

14. How many times more acidic is a solution of pH 2 than a solution of pH 4?

• The solubility of a molecular substance in water depends on the relative strength of attractive forces between solute and water molecules, compared with the strength of attractive forces between solute molecules and between water molecules.

15. Would you select ethanol, water, or lamp oil to dissolve a nonpolar molecular substance? Explain.

16. Explain the phrase "Like dissolves like."

17. Explain why greasy dishes cannot be satisfactorily cleaned with pure water.

• The solubility of a gaseous substance in water depends on the temperature of the water and the external pressure of the gas.

18. As scuba divers descend, the pressure increases on the gases that they are breathing. How does the increasing pressure affect the amount of gas dissolved in their blood?

19. Given your knowledge of gas solubility, explain why a bottle of warm cola produces more "fizz" when opened than does a bottle of cold cola.

Connecting the Concepts

20. At room temperature, C_2H_6 is a gas but C_2H_5OH is a liquid. Suggest an explanation for this difference.

21. Predict the relative solubilities of C_2H_6 and C_2H_5OH in water.

22. From each of the following pairs, select the water source more likely to contain the higher concentration of dissolved oxygen. Give a reason for each choice.

   a. a river with rapids or a calm lake
   b. a lake in spring or the same lake in summer
   c. a lake containing only sunfish or a lake containing only pike

Extending the Concepts

23. Read the label on a container of baking soda. Compare it with the label on a can of baking powder. Which one contains an acid ingredient? Suggest a reason for including the acid in the mixture.

24. Describe how changes in solubility due to temperature could be used to separate two solid, water-soluble substances.

25. The continued health of an aquarium depends on the balance of the solubilities of several substances. Investigate how this balance is maintained in a freshwater aquarium.

26. The pH of rainwater is approximately 5.5. Rainwater flows into the ocean. The pH of ocean water, however, is approximately 8.7. Investigate reasons for the difference in pH.

WATER PURIFICATION AND TREATMENT

EDITORIAL
Attendance Urged at Special Council Meeting

A special town council meeting next Wednesday could result in important decisions affecting all citizens of Riverwood. The meeting will address two primary questions: Who is responsible for the fish kill? Who should pay the expenses of trucking water to Riverwood during the three-day water shutoff as well as any damages resulting from cancellation of the fishing tournament? These questions have financial consequences for all town taxpayers.

Those testifying at next week's public meeting include representatives of industry and agriculture, scientists taking part in the river-water analyses, and consulting engineers who have been studying the cause of the fish kill. Chamber of Commerce members representing Riverwood storeowners, representatives from the County Sanitation Commission, and officials of the Riverwood Taxpayer Association also will make presentations.

We urge you to attend and participate in this meeting. Many unanswered questions remain. Was the fish kill an "act of nature" or was it due to some human error? Was there negligence? Should the town's business community be compensated, at least in part, for financial losses resulting from the fish kill? If so, how should they be compensated and by whom? Who should pay for the drinking water brought to Riverwood? Can this situation be prevented in the future? If so, at what expense? Who will pay for it?

The *Riverwood News* will set aside part of its Letters to the Editor column in coming days for your comments on these questions and other matters related to the community's recent water crisis. For useful background information on water quality and treatment, we have prepared a special feature in today's paper that we think you will find useful.

WATER PURIFICATION THROUGH HYDROLOGIC CYCLE

BY RITA HIDALGO
Riverwood News Staff Reporter

The residents of Riverwood share a sense of relief that the Snake River fish-kill mystery has been satisfactorily solved. However, ensuring the quality of Riverwood's water supply is a long-term commitment.

To act wisely about water use, whether in Riverwood or in other communities, residents must know how clean the water is and how it can be brought up to the quality required. It should not take an emergency or crisis to focus attention on these concerns.

How do water-treatment methods address threats similar to those investigated in the recent fish-kill crisis? This article provides details on how natural water-purification systems work to ensure the safety of community water supplies—particularly in light of potential threats in the form of water contamination. A companion article in today's *Riverwood News* looks at municipal water-treatment methods—procedures that mimic, in part, water-purification processes found in nature's water cycle.

Until the late 1800s, Americans obtained water from local ponds, wells, and rainwater holding tanks. Wastewater and even human wastes were discarded into holes, dry wells, or leaching cesspools (pits lined with broken stone). Some wastewater was simply dumped on the ground.

By 1880, about one-quarter of U.S. urban households had flush toilets; municipal sewer systems were soon constructed. However, as recently as 1909, sewer wastes were often released without treatment into lakes and streams, from which water supplies were drawn at other locations. Many community leaders believed that natural waters would purify themselves indefinitely.

Waterborne diseases increased as the concentration of intestinal bacteria in drinking water rose. As a result, water filtering and chlorinating of water supplies soon began. However, municipal sewage—the combined waterborne wastes of a community—remained generally untreated. Today, with larger quantities of sewage being generated and with extensive recreational use of natural waters, sewage treatment is part of every municipality's water-processing procedures.

Nature's water cycle, the hydrologic cycle, includes water-purification steps that address many potential threats to water quality. Thermal

see Water Purification, page 5

Water Purification, from page 3

energy from the Sun causes water to evaporate from oceans and other water sources, leaving behind any heavy metals, minerals, or molecular substances that were in the water.

This natural process accomplishes many of the same results as distillation. Water vapor rises, condenses into tiny droplets in clouds, and—depending on the temperature—eventually falls as rain or snow. Raindrops and snowflakes are nature's purest form of water, containing only dissolved atmospheric gases. However, human activities release a number of gases into the air, making present-day rain less pure than it used to be.

When raindrops strike soil, the rainwater collects additional impurities. Organic substances deposited by living creatures become suspended or dissolved in the rainwater. Located a few centimeters below the soil surface, bacteria feed on these substances, converting them into carbon dioxide, water, and other simple compounds. Such bacteria thus help repurify the water.

As water seeps farther into the ground, it usually passes through gravel, sand, and even rock. Waterborne bacteria and suspended matter are removed (filtered out). Thus three processes make up nature's water-purification system.

- **Evaporation,** then followed by **condensation,** removes nearly all dissolved substances.
- **Bacterial action** converts dissolved organic contaminants into a few simple compounds.
- **Filtration** through sand and gravel removes nearly all suspended matter.

Given appropriate conditions, people could depend solely on nature to purify their water. "Pure" rainwater is the best natural supply of clean water. If water seeping through the ground encountered enough bacteria for a long enough time, all natural organic contaminants could be removed. Flowing through sufficient sand and gravel would remove suspended matter from the water. However, nature's system cannot be overloaded if it is to work well.

If groundwater is slightly acidic (pH less than 7) and passes through rocks containing slightly soluble compounds such as magnesium and calcium minerals, a problem arises. Chemical reactions with these minerals may add substances to the water rather than removing them. In this case, the water may contain a relatively high concentration of dissolved minerals.

> Your first laboratory activity in this unit (page 8) demonstrated that sand can act as a water filter.

> CD-ROM WWW. **Section A: The Hydrologic Cycle**

> Recall that the water cycle was first described on page 14.

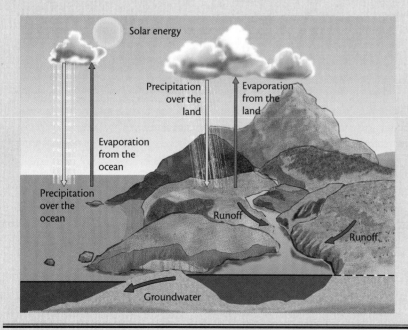

Solar energy

Precipitation over the land

Evaporation from the land

Evaporation from the ocean

Precipitation over the ocean

Runoff

Runoff

Groundwater

WATER PURIFICATION THROUGH MUNICIPAL TREATMENT

BY RITA HIDALGO
Riverwood News Staff Reporter

Today, many rivers—such as the Snake River in Riverwood—are both a source of municipal water and a place to release wastewater (sewage). Therefore, water must be cleaned twice—once before its use, and again after its use. Pre-use cleaning, often called **water treatment,** takes place at a municipal filtration and treatment plant. It is the focus of this article.

The steps in a typical water-treatment process begin when intake water flows through metal screens that pre-

vent fish, sticks, beverage containers, and other large objects from entering the water-treatment plant.

Chlorine, a powerful disinfecting agent, is added early in the water-treatment process to kill disease-causing organisms that may be present in the water. This step is called **pre-chlorination.** Then crystals of alum—aluminum sulfate,

see Municipal Treatment, page 6

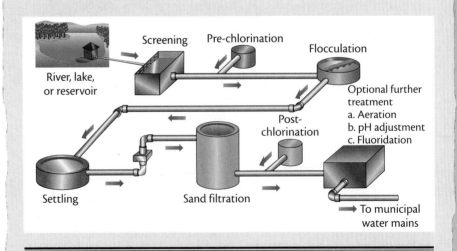

Screening Pre-chlorination Flocculation

River, lake, or reservoir

Optional further treatment
a. Aeration
b. pH adjustment
c. Fluoridation

Post-chlorination

Settling   Sand filtration

To municipal water mains

$Al_2(SO_4)_3$—and slaked lime—calcium hydroxide, $Ca(OH)_2$—are added to remove suspended particles such as colloidal clay from the water. These suspended particles can give water an unpleasant, murky appearance. The alum and slaked lime react to form aluminum hydroxide, $Al(OH)_3$, a sticky, jellylike material that traps and removes the suspended particles. This process is called **flocculation.** The aluminum hydroxide (holding trapped particles from the water) and other solids remaining in the water are allowed to settle to the tank bottom. Any remaining suspended materials that do not settle out are removed by filtering the water through sand.

In the **post-chlorination** step, the chlorine concentration in the water is adjusted to ensure that a low, but sufficient, concentration of residual chlorine remains in the water, thus protecting it from bacterial infestation.

Finally, depending on community regulations, one or more additional steps might take place before water leaves the municipal treatment plant. Sometimes water is sprayed into the air to remove odors and improve its taste. This process is known as **aeration.**

Water may sometimes be acidic enough to slowly dissolve metallic water pipes. This process not only shortens pipe life, but may also cause copper (Cu^{2+}) as well as cadmium (Cd^{2+}) and other undesirable ions to enter the water supply. Lime—calcium oxide, CaO, a basic substance—may be added to neutralize such acidic water, thus raising its pH to a proper level.

As much as about 1 ppm of fluoride ion (F^-) may be added to the treated water in a process called **fluoridation.** Even at this low concentration, fluoride ions can reduce tooth decay, as well as the number of cases of osteoporosis (bone-calcium loss) among older adults and hardening of the arteries.

WATER PURIFICATION
Building Skills 7

Refer to the two water-treatment articles by Ms. Hidalgo featured in the *Riverwood News* to answer these questions.

1. Compare natural water-treatment steps to the treatment steps found in municipal water systems.

   a. What are key similarities?
   b. What are key differences?

2. After reading the two water-treatment articles, a Riverwood resident wrote a letter to the *Riverwood News* proposing that the town's water-treatment plant be shut down. The reader pointed out that this would save taxpayers considerable money because "natural water treatment can take care of our needs just as well." Do you support the reader's proposal? Explain your answer.

Chlorine is probably the best known and most common substance used for water-treatment. It is found not only in community water supplies, but also in swimming pools. The following *Riverwood News* article provides background on chlorine's role in water treatment.

CHLORINE IN PUBLIC WATER SUPPLIES

BY RITA HIDALGO

Riverwood News Staff Reporter

The single most common cause of human illness throughout the world is unhealthful water supplies. Without a doubt, chlorine added to public water supplies has helped to save countless lives by controlling waterborne diseases. When added to water, chlorine kills disease-producing microorganisms.

In most municipal water-treatment systems, **chlorination,** or chlorine addition, takes place in one of three ways.

- *Chlorine gas, Cl_2, is bubbled into the water.* Chlorine gas, which is a nonpolar substance, is not very soluble in water. It does react with water, however, to produce a water-soluble, chlorine-containing compound.

- *A water solution of sodium hypochlorite, NaOCl, is added to the water.*

- *Calcium hypochlorite, $Ca(OCl)_2$, is dissolved in the water.* Available as both a powder and small pellets, calcium hypochlorite is often used in swimming pools. It is also a component of some solid products sold as bleaching powder.

Regardless of how chlorination takes place, chemists believe that chlorine's most active form in water is hypochlorous acid, HOCl. This substance forms whenever chlorine, sodium hypochlorite, or calcium hypochlorite dissolves in water.

Unfortunately, there is a potential problem associated with adding chlorine to municipal water. Under some conditions, chlorine in water can react with organic compounds produced by decomposing animal and plant matter to form substances harmful to human health.

One group of such substances is known as the trihalomethanes (THMs). A common THM is chloroform, $CHCl_3$, which is carcinogenic, or cancer causing.

Because of concern about the possible health risks associated with THMs, the Environmental Protection Agency has placed a current limit of 100 ppb (parts per billion) on total THM concentration in municipal water-supply systems.

Possible risks associated with THMs must be balanced, of course, against the disease-prevention benefits of chlorinated water.

Laundry bleach is a sodium hypochlorite solution.

D.4 CHLORINATION AND THMs

Several options are available to operators of municipal water-treatment plants that would help eliminate the possible THM health risks highlighted in the newspaper article that you just read. However, each method has its disadvantages.

- ◆ Treatment-plant water can be passed through an activated charcoal filter. Activated charcoal can remove most organic compounds from water, including THMs. *Disadvantage:* Charcoal filters are expensive to install and operate. Disposal also poses a problem because used filters cannot be easily cleaned of contamination. They must be replaced relatively often.

- ◆ Chlorine can be completely eliminated. Ozone (O_3) or ultraviolet light could be used to disinfect the water. *Disadvantage:* Neither ozone nor ultraviolet light protects the water once it leaves the treatment plant. Treated water can be infected by the subsequent addition of bacteria—through faulty water pipes, for example. In addition, ozone can pose toxic hazards if not handled and used properly.

- ◆ Pre-chlorination can be eliminated. Chlorine would be added only once, after the water has been filtered and much of the organic material removed. *Disadvantage:* The chlorine added in post-chlorination can still promote the formation of THMs, even if to a lesser extent than with pre-chlorination. Additionally, a decrease in chlorine concentration might allow bacterial growth in the water.

Your teacher will divide the class into working groups. Your group will be responsible for one of the three options just outlined. Answer the following questions.

1. Consider the alternative assigned to your group. Is this choice preferable to standard chlorination procedures? Explain your reasoning.

2. Can you suggest alternatives other than the three given here?

BOTTLED WATER VERSUS TAP WATER

ChemQuandary 3

When people do not like the taste of tap water, think it is unsafe to drink, or do not have access to other sources of fresh water, they may go to a vending machine or a market to buy bottled water. This bottled water often comes from a natural source, such as a mountain spring, or it may be processed at the bottling plant.

Is this water any better for you than tap water? Could it actually be more harmful? What determines water quality, and how can the risks and benefits of drinking water from various sources be assessed?

As usual, answering challenging questions requires gathering reliable data and information, weighing alternatives, and making informed decisions. Working with a partner, answer the following questions:

1. In your view, what are the most important two or three factors or considerations to analyze in deciding whether to drink tap water or bottled water?

2. For each factor listed in your answer to Question 1, what factual information would you need to gather to establish the advantages and disadvantages of drinking bottled water rather than tap water?

D.5 WATER SOFTENING

Introduction

Water containing an excess of calcium (Ca^{2+}), magnesium (Mg^{2+}), or iron(III) (Fe^{3+}) ions is called **hard water.** Hard water does not form a soapy lather easily. River water usually contains low concentrations of hard-water ions. However, as groundwater flows over limestone, chalk, and other minerals containing calcium, magnesium, and iron, it often gains higher concentrations of these ions, thus producing hard water. See Figures 31 and 32.

In this laboratory activity, you will explore several ways of softening water by comparing the effectiveness of three water treatments for

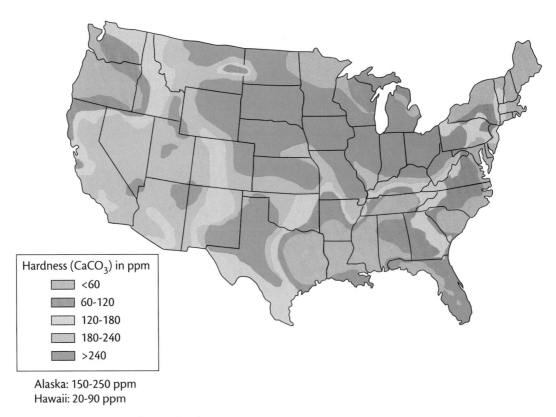

Hardness ($CaCO_3$) in ppm

	<60
	60-120
	120-180
	180-240
	>240

Alaska: 150-250 ppm
Hawaii: 20-90 ppm

Figure 31 *U.S. groundwater hardness.*

Some Minerals Contributing to Water Hardness		
Mineral	**Chemical Composition**	**Chemical Formula**
Limestone or chalk	Calcium carbonate	$CaCO_3$
Magnesite	Magnesium carbonate	$MgCO_3$
Gypsum	Calcium sulfate	$CaSO_4 \cdot 2H_2O$
Dolomite	Calcium carbonate and magnesium carbonate combination	$CaCO_3 \cdot MgCO_3$

Figure 32 *The names, chemical composition, and chemical formulas for several minerals that contribute to water hardness.*

The "hardness" of water samples can be classified as follows:
Soft
   < 120 ppm $CaCO_3$
Moderately hard
   120–350 ppm $CaCO_3$
Very hard
   > 350 ppm $CaCO_3$

removing calcium ions from a hard-water sample: sand filtration, treatment with Calgon, and treatment with an ion-exchange resin.

Calgon (which contains sodium hexametaphosphate, $Na_6P_6O_{18}$) and similar commercial products "remove" hard-water cations by causing them to become part of larger soluble anions. For example:

$$2\ Ca^{2+}(aq) \quad + \quad (P_6O_{18})^{6-}(aq) \quad \longrightarrow \quad [Ca_2(P_6O_{18})]^{2-}(aq)$$

Calcium ion   Hexametaphosphate   Calcium hexametaphosphate
from hard water   ion   ion

Calgon also contains sodium carbonate, Na_2CO_3, which softens water by removing hard-water cations as precipitates such as calcium carbonate, $CaCO_3$. The equation for the reaction is shown below. Solid calcium carbonate particles are washed away with the rinse water.

$$Ca^{2+}(aq) \quad + \quad CO_3{}^{2-}(aq) \quad \longrightarrow \quad CaCO_3(s)$$

Calcium ion   Carbonate ion   Calcium carbonate
from hard   from sodium   precipitate
water   carbonate

Another water-softening method relies on **ion exchange.** Hard water is passed through an ion-exchange resin such as those found in home water-softening units. The resin consists of millions of tiny, insoluble, porous beads capable of attracting cations. Cations causing water hardness are retained on the ion-exchange resin; cations that do not cause hardness (often Na^+) are released from the resin into the water to take the place of those that do. You will learn more about water-softening procedures after you have completed this laboratory activity.

Two laboratory tests will help you decide whether your hard-water sample has been softened. The first test is the reaction between hard-water calcium cations and carbonate anions (added as sodium carbonate, Na_2CO_3, solution) to form a calcium carbonate precipitate. The equation for this reaction is shown above. The second test is to observe the effect of adding soap to the water sample to form a lather.

Procedure

1. In your laboratory notebook, prepare a data table similar to the one shown on page 78.

Test	Filter Paper	Filter Paper and Sand	Filter Paper and Calgon	Filter Paper and Ion-Exchange Resin
Reaction with sodium carbonate (Na$_2$CO$_3$)				
Degree of cloudiness (turbidity) with Ivory soap				
Height of suds				

SAMPLE FOR REFERENCE ONLY

2. Prepare the equipment as shown in Figure 33. Lower the tip of each funnel stem into a test tube supported in a test-tube rack.

3. Fold four pieces of filter paper; insert one in each funnel. (In the following steps, funnels will be referenced, from left to right, as 1 to 4.)

4. Funnel 1 should contain only the filter paper; it serves as the control. (Hard-water ions in solution cannot be removed by filter paper.) Fill Funnel 2 one-third full of sand. Fill Funnel 3 one-third full of Calgon. Fill Funnel 4 one-third full of ion-exchange resin.

5. Pour about 5 mL of hard water into each funnel. Do not pour any water over the top of the filter paper or between the filter paper and the funnel wall.

6. Collect the filtrates in the test tubes. NOTE: The Calgon filtrate may appear blue because of other additives in the softener. They will not cause a problem. But if the filtrate appears cloudy, which means that some Calgon powder passed through the filter paper, use a new piece of filter paper and refilter the test-tube liquid.

7. Add 10 drops of sodium carbonate (Na$_2$CO$_3$) solution to each filtrate. Does a precipitate form? Record your observations. A cloudy precipitate indicates that the Ca^{2+} ion (a hard-water cation) was not removed.

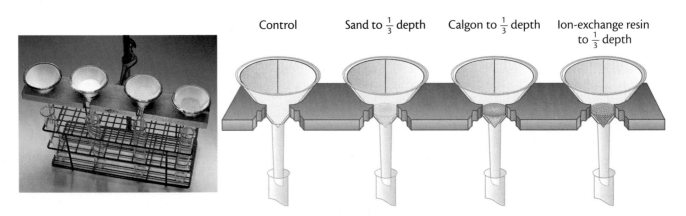

Control   Sand to $\frac{1}{3}$ depth   Calgon to $\frac{1}{3}$ depth   Ion-exchange resin to $\frac{1}{3}$ depth

Figure 33 *Filtration setup.*

8. Discard the test-tube solutions. Clean the test tubes thoroughly with tap water and rinse with distilled water. Place the test tubes back in the test-tube rack as in Step 2. Do not empty or clean the funnels; they will be used in the next step.

9. Pour another 5 mL of hard-water sample through each funnel, collecting the filtrates in the clean test tubes. Adjust test tube liquid heights, if necessary, to make all filtrate volumes equal.

10. Add one drop of Ivory liquid hand soap (not liquid detergent) to each test tube.

11. Stir each test tube gently. Wipe the stirring rod before inserting it into another test tube.

12. Compare the cloudiness, or **turbidity,** of the four soap solutions. Record your observations. The greater the turbidity, the greater the quantity of soap that dispersed. The quantity of dispersed soap determines the cleaning effectiveness of the solution.

13. Stopper each test tube and then shake vigorously, as demonstrated by your teacher. The more suds that form, the softer the water. Measure the height of suds in each test tube and record your observations.

14. Wash your hands thoroughly before leaving the laboratory.

Questions

1. Which was the most effective water-softening method? Suggest why this method worked best.

2. What relationship can you describe between the amount of hard-water ion (Ca^{2+}) remaining in the filtrate and the dispersion (cloudiness) of Ivory liquid hand soap?

3. What effect does this relationship have on the cleansing action of the soap?

4. Explain the advertising claim that Calgon prevents "bathtub ring." Base your answer on observations made in this laboratory activity.

D.6 WATER AND WATER SOFTENING

Hard water causes some common household problems. It interferes with the cleaning action of soap. As you observed, when soap mixes with soft water, it disperses to form a cloudy solution topped with a sudsy layer. In hard water, however, soap reacts with hard-water ions to form insoluble compounds (precipitates). These insoluble compounds appear as solid flakes or a sticky scum—the source of a bathtub ring. The precipitated soap is no longer available for cleaning. Worse yet, soap curd can deposit on clothes, skin, and hair. The structural formula for this objectionable substance, the product of the reaction of soap with calcium ions, is shown in Figure 34 on page 80.

If hydrogen carbonate (bicarbonate, HCO_3^-) ions are present in hard water, boiling the water causes solid calcium carbonate ($CaCO_3$) to form. The reaction removes undesirable calcium ions and thus softens the water.

Figure 34 *Structural formula of a typical soap scum. This substance is calcium stearate.*

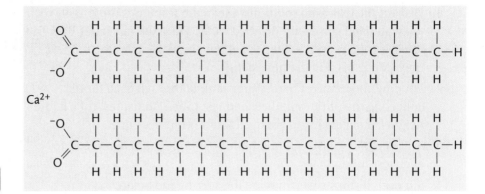

Hard Water & Soap Scum

However, the solid calcium carbonate produces rocklike scale inside tea kettles and household hot-water heaters. This scale (the same compound found in marble and limestone) acts as thermal insulation, partly blocking heat flow to the water. More time and energy are required to heat the water. Such deposits can also form in home water pipes. In older homes with this problem, water flow can be greatly reduced.

Fortunately, it is possible to soften hard water by removing some calcium, magnesium, or iron(III) ions. Adding sodium carbonate to hard water (as you did in the preceding laboratory activity) was an early method of softening water. Sodium carbonate (Na_2CO_3), known as washing soda, was commonly added to laundry water along with the clothes and soap. Hard-water ions, precipitated as calcium carbonate ($CaCO_3$) and magnesium carbonate ($MgCO_3$), were washed away in the rinse water. Water softeners in common use today include borax, trisodium phosphate, and sodium hexametaphosphate (Calgon). As you learned, Calgon does not tie up hard-water ions as a precipitate, but rather as a new, soluble ion that does not react with soap.

Figure 35 *In the 1960s and early 1970s, detergent molecules that were not decomposed by microorganisms caused some waterways to fill with sudsy foam, such as shown here. The development of biodegradable detergent molecules put an end to this unusual kind of water pollution.*

Most cleaning products sold today contain detergents rather than soap. Synthetic detergents act like soap but do not form insoluble compounds with hard-water ions. Unfortunately, many early detergents were not easily decomposed by bacteria in the environment—that is, they were not biodegradable. At times, "mountains" of foamy suds were observed in natural waterways. See Figure 35. These early detergents also contained phosphate ions (PO_4^{3-}) that encouraged extensive algae growth, choking lakes and streams. Because most of today's detergents are biodegradable and phosphate free, they do not cause these problems.

If you live in a hard-water region, your home plumbing may include a **water softener.** Hard water flows through a tank containing an ion-exchange resin similar to the one that you used in the water-softening laboratory activity. Initially, the resin is filled with sodium cations (Na^+). Calcium and magnesium cations in the hard water are attracted to the resin and become attached to it. At the same time, sodium cations leave the resin and dissolve in the water. Thus undesirable hard-water ions are exchanged for sodium ions, which do not react to form soap curd or water-pipe scale. Figure 36 illustrates this process.

It may not be economical to soften water of hardness less than about 200 ppm $CaCO_3$.

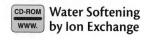 **Water Softening by Ion Exchange**

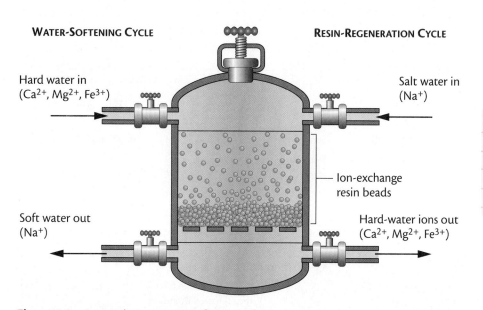

| WATER-SOFTENING CYCLE | RESIN-REGENERATION CYCLE |

Hard water in (Ca^{2+}, Mg^{2+}, Fe^{3+})

Salt water in (Na^+)

Ion-exchange resin beads

Soft water out (Na^+)

Hard-water ions out (Ca^{2+}, Mg^{2+}, Fe^{3+})

Figure 36 *Ion-exchange water-softener cycles.*

A water-softening unit typically uses from 5 to 6 pounds (2 to 3 kg) of sodium chloride (salt) for one regeneration.

As you might imagine, the resin eventually fills with hard-water ions and must be **regenerated.** Concentrated salt water (containing sodium ions and chloride ions) flows through the resin, replacing the hard-water ions held on the resin with sodium ions. Released hard-water ions wash down the drain with excess salt water. Because this process takes several hours, it is usually completed at night. After the resin has been regenerated, the softener is again ready to exchange ions with incoming hard water.

Water softeners are most often installed in individual homes. Other water treatment is done at a municipal level, both in Riverwood and in other cities.

 **Questions & Answers**

Purifying Water Means More Than Going with the Flow

When you drink from a water fountain, do you ever wonder where the water comes from? In some parts of the country, drinking water is provided by people such as **Phil Noe.**

Phil is Production Manager at Island Water Association (IWA), which provides water for Florida's Sanibel and Captiva Islands. "We have a limited supply of fresh water," says Phil, "so we've built a plant that lets us use water from our aquifers." Aquifers are underground layers of permeable (porous) sand and limestone that contain large quantities of water.

> "We have a limited supply of fresh water," says Phil, "so we've built a plant that lets us use water from our aquifers."

After pumping the brackish, undrinkable water up from the Suwanee Aquifer to IWA's processing plant, Phil and his co-workers remove almost all of the salt and minerals, producing water that is more pure than many mountain streams. The process that accomplishes this task is known as reverse osmosis. The accompanying illustration is a comparison of osmosis with reverse osmosis. The amount of pressure applied in reverse osmosis must exceed the osmotic pressure, which tends to move the water from a region of higher vapor pressure to one of lower vapor pressure. Because osmotic pressure depends only on the concentration of the salt solution, it is known as a colligative property.

Aquifers are located at depths ranging between 700 and 900 feet. Water pumped up from these aquifers contains an average of 3000 ppm total dissolved solids (TDS). Feed water is pumped through hundreds of feet of pipes containing salt-filtering membranes. In the process, the purified water, called the permeate, separates from the salts and other dissolved solids. Eventually, the water is purified to levels as low as 100 ppm TDS.

Because Florida is subject to hurricanes and other tropical storms, Sanibel Island's water system maintains

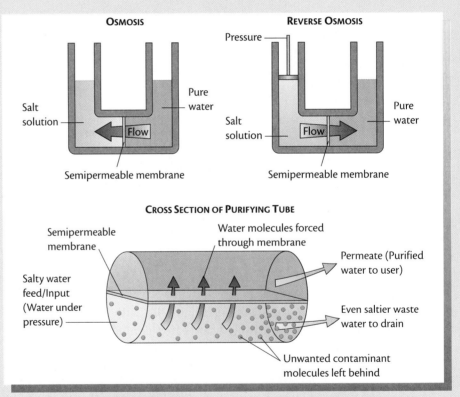

Osmosis occurs naturally when water in a dilute solution passes through a semipermeable membrane into a concentrated solution. In reverse osmosis, pressure must be applied to a concentrated solution to force water through a semipermeable membrane into a dilute solution.

After filtering to remove the larger solids, the water passes through equipment where reverse osmosis is used to produce purified water.

Saving up for a windy day.

15 million gallons of purified water in storage. As a result, island customers can manage without the plant in operation during brief periods of severe weather with no more inconvenience than several weeks without lawn sprinklers.

Eighty percent of IWA's feed water is converted into drinkable water. The remaining 20% is pumped out into the Gulf of Mexico. This water, a salt solution, is lower in salt and mineral concentrations than the gulf water itself and is harmless to marine life.

Web Search . . .

◆ Use the World Wide Web to search for other communities that rely on reverse osmosis for all or part of their water supply.

◆ Most communities use rivers, lakes, or wells to supply their water needs. Search the Web for information about municipal water-supply systems that use alternative water-purification processes in addition to reverse osmosis.

◆ What other kinds of water purification systems are in use in the United States?

◆ Use the Web to gather information about the cost of water for household use in various parts of the United States. Is there a connection between the cost of water and the location of the community? For example, is water more expensive in the desert than it is near the ocean?

Aquatic wildlife is unaffected by the salt water left over from the reverse osmosis process.

SECTION SUMMARY

Reviewing the Concepts

♦ **Water can be purified through the actions of the hydrologic cycle or through municipal treatment.**

1. Explain how water can be purified through the actions of the hydrologic cycle.

2. How are the properties of aluminum hydroxide related to the process of flocculation?

3. Why is calcium oxide (CaO) sometimes added in the final steps of municipal water treatment?

♦ **Chlorination is commonly used to treat and purify water for human consumption. The benefits of chlorine treatment can be weighed against its risks.**

4. What advantages does chlorinated drinking water have over untreated water?

5. What are some disadvantages of using chlorinated water?

6. Water from a clear mountain stream may require chlorination to make it safe for drinking. Explain why.

♦ **Hard water contains relatively high concentrations of calcium, magnesium, or iron cations. Such water can be softened by removing most of these ions as precipitates or complexes.**

7. What is the origin of the expression "hard water"?

8. When a sample of well water was mixed with a few drops of sodium carbonate solution, a precipitate formed. What does the formation of a precipitate indicate about the water sample?

9. Hard water often tastes better than distilled water. Explain.

10. Which source in a given locality would probably have harder water, a well or a river? Explain.

Connecting the Concepts

11. A simple test for the hardness of water is to add soap to the sample and shake it. Explain how measuring the quantity of soap suds formed can be used to assess the hardness of water.

12. Explain how hard water can interfere with the processes of a water-treatment plant.

Extending the Concepts

13. How much water would you have to drink to get your minimum daily requirement of calcium from water that contains 300 ppm calcium carbonate?

14. Explain why hard water stains in old sinks are found more often around hot-water faucets than around cold-water faucets.

15. Investigate the reasons why sodium-based ion-exchange resins have been banned in some municipalities.

PUTTING IT ALL TOGETHER

FISH KILL—FINDING THE SOLUTION

FISH KILL CAUSE FOUND

MEETING TONIGHT

BY ORLANI O'BRIEN
Riverwood News Staff Reporter

Mayor Edward Cisko announced at a news conference held early today that the cause of the fish kill in the Snake River has been determined. The details of the accidental cause will be released at a town council meeting tonight. As of today, levels of all dissolved materials in the river are normal and the water should be considered safe.

Accompanying Mayor Cisko was Dr. Harold Schmidt of the Environmental Protection Agency. Dr. Schmidt performed dissections on fish taken from the river within a few hours of their death. He also directed the team that analyzed accumulated river-water data in efforts that led to determining the cause of the fish kill. Dr. Schmidt gave assurances that the town's water supply is "fully safe to drink."

Mayor Cisko refused to elaborate on reasons for the accident. However, he invited the public to the special town council meeting scheduled for 8 P.M. tonight at Town Hall. The council will discuss events that caused the fish kill and how costs associated with the three-day water shutoff will be paid. Several area groups, as well as invited experts, plan to make presentations at tonight's meeting.

TOWN COUNCIL MEETING

Meeting Rules and Penalties for Rule Violations

 **Putting It All Together**

1. The order of presentations is decided by council members and announced at the start of the meeting.

2. Each group will have a specified time for its presentation. Time cards will notify the speaker of time remaining.

3. If a member of another group interrupts a presentation, the offending group will be penalized 30 seconds for each interruption, to a maximum of one minute. If the group has already made its presentation, it will forfeit its rebuttal time.

Figure 37 *View of a city council meeting in session in Austin, Texas. Open meetings such as this encourage citizen involvement in discussions of local issues and in related decision-making challenges.*

Town Council Members: Background Information

Your group is responsible for conducting the meeting in an orderly manner. Be prepared to:

1. Decide and announce the order of presentations at the meeting. Groups presenting factual information should be heard before groups voicing opinions.

2. Decide how the presentation area will be organized: where town council members controlling the meeting will be located, where the presenters will present, and where the groups and observers will be positioned.

3. Explain the meeting rules and the penalties for violating those rules.

4. Recognize each group at its assigned presentation time.

5. Enforce established presentation time limits by preparing time cards with "one minute," "30 seconds," and "time is up" written on them. These cards, placed in the speaker's line of sight, can serve as useful warnings.

6. Control the rebuttal discussions and open-forum speeches.

7. Summarize the options when testimony has been completed.

8. Conduct a vote of all town council members.

9. Report the results of the vote and future actions mandated by it.

Power Company Officials: Background Information

The power plant includes a dam and reservoir that ensure an adequate supply of cooling water. Your company's engineers control the rate of water release from the dam into the Snake River.

Normally, only relatively small volumes of water are released at any particular time. However, releasing large quantities of water from the dam is a standard way of preventing flooding. The last time such a large volume of water was released from the dam was 30 years ago. A fish kill was reported then, but the cause remained unknown. On that occasion, Riverwood and surrounding areas had experienced an unusually wet summer.

The dam, constructed in the 1930s, had the most current design of that time. Since then, its basic structure has not been modified.

Agricultural Cooperative Representatives: Background Information

Cooperative members in the Snake River valley include farmers and ranchers managing a variety of crops and livestock. Your cooperative assumes a proactive role in informing its members of current best practices and regulations regarding the use of agrochemicals and the management of wastes and runoff from fields and pastures.

Heavy rains present a problem for farmers. Although the rain is good for crops, it can wash away recently applied fertilizers and pesticides. This is not only expensive, but it can cause problems if these substances wash into the watershed.

Mining Company Representatives: Background Information

Riverwood began as a mining town on the Snake River, which provided early residents with a source of water. Your company intensely mined the hills surrounding Riverwood in the 1930s and 1940s. The important metals that came from this area included zinc and silver. The by-products of mining and processing the metal ores were collected in storage ponds built in accordance with the specifications and regulations of that time.

In seasons with average rainfall, the runoff from the waste ponds contains heavy-metal ions at levels within the values specified by your company's EPA permit. Your company monitors effluent values and keeps the ponds secured. Your company's structural engineers are responsible for upkeep of storage ponds at abandoned mine sites. However, during heavy rainfall, some underground settling in the mines and avalanches in hilly areas of the Snake River have been noted.

Scientists: Background Information

You are responsible for explaining how the analyzed data support the proposed cause of the fish kill. You should be prepared to explain what the data mean and why data fluctuations are noted from month to month or year to year. You may be called on to explain concepts such as pH, solubility of molecular and ionic substances, units of concentration, water-purification techniques, the hydrologic cycle, and other water-related concepts. It is important that you help council members and other attendees understand how the analyzed data document the cause of the fish kill.

Consulting Engineers: Background Information

Your consulting firm was hired to do a detailed examination of the cause of the fish kill. Your task was to determine whether accident, human error, negligence, or an unforeseen circumstance was responsible for the Riverwood crisis. In addition, you were asked to prepare scenarios or suggest improvements that would prevent recurring fish kills.

Your presentation should include the proposed solutions, the costs and benefits of each solution, and a detailed analysis of the cause. It is understood that you may not be familiar with cost analyses of major projects; however, you should try to make feasible estimates.

Chamber of Commerce Members: Background Information

Canceling the annual fishing tournament cost you and other Riverwood merchants a substantial sum of money. Close to one thousand out-of-town tournament participants were expected. Many would have rented rooms for at least two nights and eaten at local restaurants and fast-food establishments. In anticipation of this business, extra food supplies and support services were ordered. Fishing and sporting goods stores stockpiled extra fishing supplies. Some businesses have applied for short-term loans to help pay for their unsold inventories.

Local churches and the high school planned family social activities as revenue makers during the tournament weekend. For example, the school band scheduled a benefit concert that would have raised money to send band members to the spring band competition.

People are likely to remember the fish kill for many years. Tournament organizers predict that future fishing competition revenues in Riverwood will be substantially reduced due to this year's adverse publicity. Thus total financial losses resulting from the water emergency may be much higher than most current estimates predict.

You should be able to discuss how merchants and businesses were affected by this event and summarize the availability of support (as well as lack of support) to help resolve the issues.

County Sanitation Commission Members: Background Information

You are responsible for the protection and safety of the Snake River water supply. You are the group that completes most of the routine water testing for the supply of drinking water in Riverwood. It is important to know what the standards that specify the quality of drinking water mean and to explain how the water testing is done. You should be able to report maximum contamination levels (MCLs) for hazardous water contaminants. You should also know the allowable limits or expected ranges for other analyzed water data.

Riverwood Taxpayer Association Members: Background Information

Your organization is concerned about the financial effects of the fish kill on Riverwood citizens. Thus some of the important questions to be answered at the town council meeting should be addressed in your presentation. These questions include:

♦ Who will pay for the water brought into Riverwood during the water shutoff?

♦ Will taxes be increased to compensate local businesspeople for their financial losses? (Keep in mind that local merchants themselves are likely to be Riverwood taxpayers!)

♦ If the organization responsible for the fish kill takes measures to prevent its recurrence, will the costs be passed on to consumers? If so, how?

Because your presentation may be influenced by the testimony of other groups, you may find it useful to obtain their written briefs before the council meeting, if possible.

LOOKING BACK AND LOOKING AHEAD

The Riverwood water mystery is solved! In the end, scientific data provided the answer. And now human ingenuity will provide strategies for preventing the recurrence of such a crisis. In the course of solving the problem, the citizens of Riverwood learned about the water that they take for granted—abundant, clean water flowing steadily from their taps—and probably gained a greater appreciation for it.

Although Riverwood and its citizens exist only on the pages of this textbook, their water-quality crisis could be very real. The chemical facts, principles, and procedures that clarified their problem and its solution have applications in your own home and community.

Although the fish-kill mystery has been solved, your exploration of chemistry has only just begun. Many issues related to chemistry in the community remain, and water and its chemistry are only one part of a larger story.

UNIT 2

HOW can the chemical
and physical properties
of matter be explained?

WHERE are mineral
resources found and
how are they
processed?

MATERIALS: STRUCTURE AND USES

WHAT information do chemical equations convey about matter and its changes?

HOW can matter be modified to make it more useful?

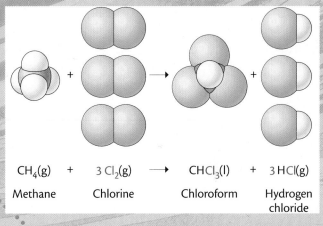

$$CH_4(g) + 3\,Cl_2(g) \longrightarrow CHCl_3(l) + 3\,HCl(g)$$

| Methane | Chlorine | Chloroform | Hydrogen chloride |

Your local Congresswoman has invited you and your classmates to submit a design for a new U.S. half-dollar coin. What will the coin's composition be? Why? What makes a particular material best suited for each intended use? Turn the page to learn the answers to these questions—and perhaps be the lucky contest winner.

MEMORANDUM

TO: District 12 High School Principals and
              Chemistry Teachers

FROM: The Hon. Maria Gonzales
              U.S. House of Representatives

SUBJECT: Coin-Design Competition

As you may know, I plan to introduce a Bill in the House of Representatives authorizing the production of a new half-dollar coin by the United States Mint. The purpose of this memorandum is to let you know about a contest my office is sponsoring for high-school chemistry students in your Congressional District. The competition involves proposing a possible design for the new coin. The winning design will be used as an example in House Committee deliberations and on the floor of the House as I seek support from colleagues for authorization of the new 50-cent coin.

Each high school within your Congressional District may submit one complete design. The team of students (or student) who creates the winning design will be honored, along with the rest of their chemistry class, at an open house at my local office. In addition, the winning student or team will travel with my staff to Washington, D.C., for the introduction of my Bill and for a public presentation of the suggested coin design.

A complete coin-design proposal must include the following information:

- Full name(s) and address(es) of the designer(s)
- Chemistry teacher's name and course title
- School name and phone number
- Coin diameter, thickness, and mass
- Detailed drawing or actual model of the coin, enlarged 5x for clarity
- Specifications for the composition of the coin's material
- Plans for obtaining or creating the materials used in the coin
- Two-page rationale for key decisions made in the coin's design

All completed proposals will be due in my office within six weeks of receipt of this memorandum.

MG/hs

Buoyed by the U.S. Mint's release of a new dollar coin and the success of the recent quarter-coin series featuring each state, Congressional Representative Maria Gonzales is planning to introduce a Bill authorizing the U.S. Mint to produce a new half-dollar coin.

Realizing that Congressional colleagues will request information about the proposed replacement, Representative Gonzales is sponsoring a contest for local high school chemistry students to propose a prototype design of the new coin.

As her memo suggests, every aspect of the half-dollar coin—from its appearance to its size and composition—is to be included in the designs. Because only one design can be submitted from each school, your class will have to decide which team's coin proposal will be selected for further screening in this competition.

As you go on to consider important features in your coin's design, you will learn about Earth's mineral resources and how they are used by society. You will learn why certain materials are used for particular new products—coins and other useful things—and how those materials are developed from available resources. Throughout this unit, keep in mind how such chemical knowledge can help guide your design of a new coin.

WHY WE USE WHAT WE DO

Every human-produced object, old or new, is made of materials selected for their specific characteristics, or properties. What makes a particular material best for a particular use? You can begin to answer this question by exploring some properties of materials.

A.1 PROPERTIES MAKE THE DIFFERENCE

Overview

In this unit, you will be considering the design of something you use every day—money in the form of coins. Throughout history, people have used many different items as money: beads, stones, printed paper, even precious metals, to name a few. What characteristics make an object suitable (useful) for manufacturing money or minting coins? How important is appearance? Size? What other important characteristics can you suggest?

As you already know, every substance has characteristic properties that distinguish it from other substances, thus allowing it to be identified. These characteristic properties include **physical properties,** or properties that can be determined without altering the chemical makeup of the material. Color, density, and odor are examples of physical properties. The ability of a material to undergo **physical changes,** such as melting, boiling, and bending, is often important in its use. Remember that in a physical change, the identity of the substance remains the same.

The **chemical properties** of a material often play important roles in its usefulness. Chemical properties relate to the tendency of a substance to undergo **chemical changes**—that is, to transform into new substances. Consider the common chemical change of rust forming on iron surfaces. The tendency of a material to rust, or react to form an oxide, is the chemical property that accounts for this chemical change. A chemical change can often be detected by observing the formation of a gas or solid, a color change, a change on the surface of a solid, or a temperature change (indicating that thermal energy has been absorbed or given off). Figure 1 illustrates some physical and chemical changes involving copper.

In the following activity, you will classify some characteristics of common materials as either physical or chemical properties.

PHYSICAL AND CHEMICAL PROPERTIES
Building Skills 1

Consider this statement: *Copper compounds are often blue or green.* Does the statement describe a physical or chemical property? To answer the question, first think about whether a change in the identity of a substance is involved. Has the substance been chemically changed? If the answer is

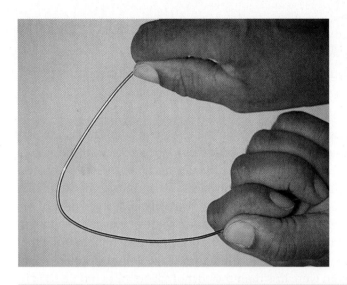

Figure 1 *Examples of physical and chemical changes involving copper.*

"no," then the statement describes a physical property; if the answer is "yes," then the description is of a chemical property.

Color is a characteristic physical property of many chemical compounds. A green copper compound in a jar on the shelf is not undergoing any change in its identity. Color, therefore, is a physical property. A change in color, however, often indicates a change in identity and thus a chemical change. For example, colored matter called litmus, derived from a plant-like organism called a lichen, turns from blue to pink when exposed to acid. This is a chemical change involving the chemical properties of litmus and acid.

Now consider this statement: *Oxygen gas supports the burning of wood.* Does the statement refer to a physical or chemical property of oxygen? If you apply the same key question—is there a change in the identity of the oxygen gas—you will arrive at the correct answer. The burning of wood— or combustion—involves chemical reactions between the wood and oxygen gas that change both reactants. The reaction products of ash, carbon dioxide, and water vapor are very different from wood and oxygen gas. Thus the statement refers to a chemical property of oxygen gas (as well as of wood).

Now it's your turn. Classify each statement as describing either a physical property or a chemical property. (*Hint:* Decide whether the chemical identity of the material does or does not change when the property is observed.)

1. Pure metals have a high luster (are shiny and reflect light). P

2. The surfaces of some metals become dull when exposed to air. C

3. Nitrogen gas, which is a relatively nonreactive element at room temperature, can form nitrogen oxides at the high temperatures of an operating automobile engine. C

4. Milk turns sour if left too long at room temperature. C

5. Diamonds are hard enough to be used on drill bits. P

6. Metals are typically ductile (can be drawn into wires). P

7. Bread dough increases in volume if it is allowed to "rise" before C baking.

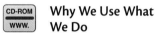

Why We Use What We Do

8. Argon gas, rather than air, is used in many lightbulbs to prevent oxidation of the metal filament. ✗ c

9. Generally, metals are better conductors of heat and electricity than are nonmetals. ρ

As you might imagine, many issues need to be considered when selecting materials for a specific use. A substance with properties well suited to a purpose may be either unavailable in sufficient quantity or too expensive. Or a substance may have undesirable physical or chemical properties that limit its use. In these and other situations, another material with most of the sought-after properties can often be found and used.

The cost of material is a particularly important issue when manufacturing coins and printed currency. Just imagine what would happen if the declared value of a coin were less than the cost of the metals contained in it. How would this affect the production and circulation of the coin? This situation actually occurred in the United States not too long ago. In the early 1980s, copper became too expensive to be used as the primary metal in pennies. In other words, the cost of the copper in a penny was greater than the value of the penny. A lower-cost replacement having similar properties was needed. Zinc, another metallic element, was chosen to replace most of the copper in all post-1982 pennies. Zinc is about as hard as copper and has a density (7.14 g/cm³) close to that of copper (8.94 g/cm³). Zinc is also readily available and less expensive than copper.

Unfortunately, zinc is more chemically reactive than copper. During World War II, copper metal was in short supply. To conserve that resource, zinc-plated steel pennies—known to coin collectors as "white cents" or "steel cents"—were created in 1943. The new pennies quickly corroded. As you can see in Figure 2, these pennies also looked considerably different from traditional copper pennies. Production of the zinc-plated pennies ended within a year.

The problems associated with the earlier zinc-plated pennies were solved in the early 1980s. In the new design, the properties of copper were used where they were most needed—on the coin's surface—and the properties of zinc where they were suitable—within the body of the coin. All post-1982 pennies are 97.5% zinc. They are composed of a zinc core surrounded by a thin layer of copper, added to increase the coin's durability and maintain its familiar appearance. Figure 3 shows a cross-section of a post-1982 penny.

Whether it is copper or zinc in a penny or tungsten in a lightbulb, the message is clear: Every substance has its own set of physical and chemical properties. But with millions of substances available, how can one identify the "best" substance(s) for a given need?

Fortunately, the list of possibilities can be greatly shortened. All substances are made of a relatively small number of building blocks—the atoms of the different chemical elements. Knowing the similarities and differences among atoms of elements and among combinations of those atoms can greatly simplify the challenge of matching a substance to its most appropriate uses.

Figure 2 *Zinc-copper and copper pennies (top); "new" and corroded zinc-steel pennies (bottom).*

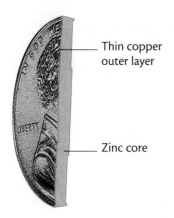

Thin copper outer layer

Zinc core

Figure 3 Cross-section showing composition of a post-1982 penny.

presenting
Atoms and
Ions

A.2 THE CHEMICAL ELEMENTS

You learned in Unit 1 that all matter is composed of atoms. One element differs from another because its atoms have properties that differ from those of all other elements. More than 100 chemical elements are known. The table in Figure 4 lists some common elements and their symbols. An alphabetical list of all the elements (names and symbols) can be found on page 108.

Elements can be grouped, or classified, in several ways according to similarities and differences in their properties. Two major classes are **metals** and **nonmetals**. Metals include such elements as iron, tin, zinc, and copper. Carbon and oxygen are examples of nonmetals. Everyday experience has given you some knowledge of metallic properties. The laboratory activity you will soon do will give you an opportunity to explore the properties of metals and nonmetals in more depth.

A relatively few elements called **metalloids** have properties that are intermediate to those of metals and nonmetals. That is, metalloids are like metals when it comes to some properties and like nonmetals when it comes to others. Examples of metalloids include silicon and germanium, both commonly used in the computer industry.

What properties of matter are used to distinguish metals, nonmetals, and metalloids? The next activity will help you find out.

A.3 METAL OR NONMETAL? : **Laboratory Activity**

Introduction

In this activity you will investigate several properties of seven elements and then decide whether each element is a metal, nonmetal, or metalloid. You will examine the color, luster, and form of each element and attempt to crush each sample with a hammer. You or your teacher (as a demonstration) will also test the substance's ability to conduct electricity. Finally, you will determine the reactivity of each element with two solutions: hydrochloric acid, HCl(aq), and copper(II) chloride, $CuCl_2$(aq).

Common Elements	
Name	**Symbol**
Aluminum	Al
Antimony	Sb
Argon	Ar
Barium	Ba
Beryllium	Be
Bismuth	Bi
Boron	B
Bromine	Br
Cadmium	Cd
Calcium	Ca
Carbon	C
Cesium	Cs
Chlorine	Cl
Chromium	Cr
Cobalt	Co
Copper	Cu
Fluorine	F
Gold	Au
Helium	He
Hydrogen	H
Iodine	I
Iron	Fe
Krypton	Kr
Lead	Pb
Lithium	Li
Magnesium	Mg
Manganese	Mn
Mercury	Hg
Neon	Ne
Nickel	Ni
Nitrogen	N
Oxygen	O
Phosphorus	P
Platinum	Pt
Potassium	K
Silicon	Si
Silver	Ag
Sodium	Na
Sulfur	S
Tin	Sn
Tungsten	W
Uranium	U
Zinc	Zn

Figure 4 A table of common elements and their symbols.

Procedure

1. In your laboratory notebook, prepare a data table that has enough space to record the properties of the seven element samples, which have been coded with letters *a* to *g*.

DATA TABLE

Element	Appearance	Conductivity (optional)	Result of Crushing	Reaction with Acid	Reaction with $CuCl_2(aq)$
a.					
b.					
c.					
d.					
e.					
f.					
g.					

2. *Appearance:* Observe and record the appearance of each element, including physical properties such as color, luster, and form. You can record the form as crystalline (like table salt), noncrystalline (like baking soda), or metallic (like iron).

3. *Conductivity:* If an electrical conductivity apparatus is available, use it to test each sample. **CAUTION:** *Avoid touching the bare electrode tips with your hands; some may deliver an uncomfortable electric shock.* Touch both electrodes to the element sample, but do not allow the electrodes to touch each other. See Figure 5. If the bulb lights, even dimly, electricity is flowing through the sample. Such a material is called a **conductor.** If the bulb fails to light, the material is a **nonconductor.**

Figure 5   *Testing for conductivity.*

4. *Crushing:* Gently tap each element sample with a hammer. Based upon the results, decide whether the sample is **malleable,** which means it flattens without shattering when struck, or **brittle,** which means it shatters into pieces.

5. *Reactivity with acid.*

   a. Label seven wells of a clean wellplate *a* through *g.*

   b. Place a sample of each element in its appropriate well. The solid wire or ribbon samples provided by your teacher will be less than 1 cm in length. Other samples should be between 0.2 g and 0.4 g—you can estimate that as no larger than the size of a match head.

   c. Add 15 to 20 drops of 0.5 M HCl to each well that contains a sample. **CAUTION:** *0.5 M hydrochloric acid (HCl) can chemically attack skin if allowed to remain in contact for a long time. If any hydrochloric acid accidentally spills on you, ask a classmate to notify your teacher immediately. Wash the affected area immediately with tap water and continue for several minutes.*

   d. Observe and record each result. The formation of gas bubbles indicates that a chemical reaction has occurred. A change in the appearance of an element sample may also indicate a chemical reaction. Decide which elements reacted with the hydrochloric acid and which did not. Record these results.

   e. Discard the wellplate contents as instructed by your teacher.

6. *Reactivity with copper(II) chloride.*

   a. Repeat Steps 5a and 5b.

   b. Add 15 to 20 drops of 0.1 M copper(II) chloride ($CuCl_2$) to each sample.

   c. Observe each system for three to five minutes—changes may be slow. Decide which elements reacted with the copper(II) chloride and which did not. Recall the criteria you used in the acid test to determine if a reaction occurred. Record each result.

   d. Discard the wellplate contents as instructed by your teacher.

7. Wash your hands thoroughly before leaving the laboratory.

Questions

1. Classify each property tested in this activity as either a physical property or a chemical property.

2. Sort the seven coded elements into two groups based on similarities in their physical and chemical properties.

3. Which element(s) could fit into either group? Why?

4. Using the following information, classify each tested element as a metal, nonmetal, or metalloid.

   ◆ Metals have a luster, are malleable, and conduct electricity.

   ◆ Many metals react with acids; many metals also react with copper(II) chloride solution.

   ◆ Nonmetals are usually dull in appearance, are brittle, and do not conduct electricity.

   ◆ Metalloids have some properties of both metals and nonmetals.

You have learned one classification scheme for elements: metals, non-metals, and metalloids. The quantity of detailed information about the elements is enormous, however. And when choosing or designing materials for specific uses, the more information available about the elements (including similarities and differences among them), the better the decisions. How then is all the knowledge about the elements conveniently organized? You have already been introduced to the answer—the periodic table. Now you will explore its origins and gain greater understanding of the chemical information it conveys.

A.4 THE PERIODIC TABLE

By the mid-1800s, about 60 elements had been identified. Five of these were nonmetals that were gases at room temperature: hydrogen (H), oxygen (O), nitrogen (N), fluorine (F), and chlorine (Cl). Two liquid elements were also known, the metal mercury (Hg) and the nonmetal bromine (Br). The rest of the known elements were solids with widely differing properties.

In an effort to impose some organization on the information related to the elements, several scientists tried to place elements with similar properties near one another on a chart. Such an arrangement is called a **periodic table**. Dimitri Mendeleev, a Russian chemist, published a periodic table in 1869. A similar table is still used today. In some respects, the periodic table has a pattern that resembles a monthly calendar, in which weeks repeat on a regular (periodic) seven-day cycle.

The periodic tables of the 1800s were organized according to two characteristics of elements. It was known that atoms of different elements have different masses. For example, hydrogen atoms have the lowest mass, oxygen atoms are about 16 times more massive than hydrogen atoms, and sulfur atoms are about twice as massive as oxygen atoms (making them about 32 times more massive than hydrogen atoms). Based on such comparisons, an **atomic weight** was assigned to each element in Mendeleev's periodic table. This atomic weight then became one of the two criteria for arranging elements in the periodic table.

The other criterion for organizing elements in early periodic tables was their respective "combining capacity" with other elements, such as chlorine and oxygen. It was known that atoms of various elements differ in the way that they combine with another element. For example, one atom of potassium (K) or cesium (Cs) combines with only one atom of chlorine (Cl) to produce the compound KCl or CsCl. Such one-to-one compounds can be represented as ECl, where E stands for the Element combining with chlorine. However, one atom of magnesium (Mg) or strontium (Sr) combines with two atoms of chlorine to produce the compound $MgCl_2$ or $SrCl_2$, which can be represented in general terms as ECl_2. Atoms of other elements may combine with three or four chlorine atoms to produce compounds with the general formula ECl_3 or ECl_4.

In the first periodic tables, elements with similar chemical properties were placed in vertical groups (columns). Horizontal arrangements were based on increasing atomic weights of the elements. In the activity that follows, you will develop a classification scheme for some elements in much the same way Mendeleev did.

A.5 GROUPING THE ELEMENTS | Making Decisions

You will be given a set of 20 element data cards. Each card lists some properties of a particular element.

1. Arrange the cards in order of increasing atomic weight.

2. Place the cards in a number of different groups. Each group should include elements with similar properties. You might need to try several methods of grouping before you find one that makes sense to you.

3. Examine the cards within each group for any patterns. Arrange the cards within each group in some logical sequence. Again, trial and error may be a useful method for accomplishing this task.

4. Observe how particular element properties vary from group to group.

5. Arrange all the card groups into some logical sequence.

6. Decide on the most reasonable and useful patterns within and among card groups. Then tape the cards onto a sheet of paper to preserve your pattern for classroom discussion.

A.6 THE PATTERN OF ATOMIC NUMBERS

Creators of early periodic tables were unable to offer explanations for similarities in properties found among neighboring elements. For example, all of the elements in the leftmost column of the periodic table are very reactive metals. All the elements in the rightmost column are unreactive gases. The reason for these similarities, which was discovered about 50 years after Mendeleev's work, serves as the basis for the modern periodic table.

As you will recall from Unit 1, all atoms are composed of smaller particles, including equal numbers of positively charged protons and negatively charged electrons. The number of protons, called the **atomic number,** distinguishes atoms of different elements. For example, every sodium atom (and only a sodium atom) contains 11 protons. The atomic number of sodium is 11. Each carbon atom contains 6 protons. If the number of protons in an atom is 9, it is a fluorine atom; if 12, it is a magnesium atom. Thus the atomic number (number of protons) identifies every atom as a particular element.

Early periodic tables, much like the one you just constructed, used atomic weights to organize the elements. Although this method produces reasonable results for elements with relatively small atomic weights, it does not work well for more massive atoms. The reason for this is the existence of another small particle that makes up the atom, the electrically uncharged neutron. The total mass of an atom is largely determined by the combined number of protons and neutrons in its **nucleus.** The nucleus is a concentrated region of positive charge (from the protons) in the center of an atom. See Figure 6.

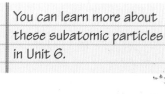

You can learn more about these subatomic particles in Unit 6.

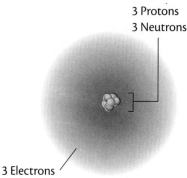

3 Protons
3 Neutrons

3 Electrons

Figure 6 *Components of a particular atom of lithium (Li).*

The total number of protons and neutrons in the nucleus of an atom is called the **mass number.** Electrons make up the rest of an atom, but because each electron is about 1/2000th the mass of a proton or neutron, the total mass of the electrons does not contribute significantly to the mass of an atom.

While all atoms of a particular element have the same number of protons, the number of neutrons can differ from atom to atom of an element. For example, carbon atoms always contain 6 protons, but they may contain 6, 7, or 8 neutrons. Thus individual carbon atoms can have mass numbers of 12, 13, or 14. Atoms with the same number of protons but different numbers of neutrons are called **isotopes.** In other words, isotopes are atoms of the same element with different mass numbers.

In the modern Periodic Table of the Elements, elements are placed in sequence according to their increasing atomic number. But because electrically neutral (uncharged) atoms contain equal numbers of protons and electrons, the Periodic Table is also sequenced by the number of electrons that the neutral atoms of each element contain.

Is there a connection between the atomic numbers used to organize the modern Periodic Table and the element properties used by nineteenth-century chemists to create their periodic tables? If there is, what is that connection? Continue reading to explore the relationship between atomic numbers and the properties of elements.

PERIODIC VARIATION IN PROPERTIES

Building Skills 2

Your teacher will assist you in identifying the atomic numbers of the 20 elements you considered earlier in the unit. Use these atomic numbers and information about each element's properties to prepare the two graphs described below. Look for patterns between atomic numbers and element properties as you construct the graphs.

Follow the graphing guidelines you learned in Unit 1 (page 66).

Graph 1. Trends in a chemical property.

1. On a sheet of graph paper, draw a set of axes and title the graph *Trends in a Chemical Property.*

2. Label the *x* axis *Atomic Number of E,* and number it from 1 to 20.

3. Label the *y* axis *Number of Oxygen Atoms per Atom of E.* Number it from 0 to 3; in increments of 0.5.

4. Construct a bar graph, as demonstrated in Figure 7, by plotting the oxide data from the element cards. For example, if no oxide forms, the height of the bar will be 0 because oxygen atoms do not form a compound with atoms of E. If E_2O (1 oxygen atom for 2 E atoms) is formed, the height of the bar is 0.5, which is the number of oxygen atoms for each E atom in the compound.

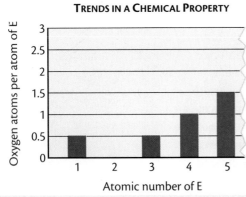

Figure 7 *Sample bar graph for oxide data.*

Similarly, the heights of the bars for other oxides are 1 for EO, 1.5 for E_2O_3, 2 for EO_2, and 2.5 for E_2O_5. Do you understand why?

5. Label each bar with the actual symbol of the element E involved in that compound.

Graph 2. Trends in a physical property.

6. On a separate sheet of graph paper, draw a set of axes, and title the graph *Trends in a Physical Property*.

7. Label the *x* axis *Atomic Number* and number it from 1 to 20.

8. Label the *y* axis *Boiling Point (K)* and number it from 0 to 3000 K, as shown in Figure 8. Use as much of the space on the graph paper as possible to plot these kelvin temperatures.

9. Construct a graph as in Step 4, this time using the boiling point data from the element cards, as shown in Figure 8. Do not include data for the element with atomic number 6. The boiling point of this element (carbon) would be quite far off the graph.

10. Label each bar with the symbol of the element it represents.

> Temperature in kelvins (K) is related to temperature in degrees Celsius (°C) by K = 273.15 + °C. You will learn more about the kelvin temperature scale in Unit 4.

QUESTIONS

1. Does either graph reveal a repeating, or cyclic, pattern? (*Hint:* Focus on elements represented by very large or very small bars.) Describe any patterns you observe.

2. Are these graphs consistent with patterns found in your earlier grouping of the elements? Explain.

3. Based on these two graphs, why is the chemist's organization of elements called a *periodic* table?

4. Where are elements with the highest oxide numbers located on the Periodic Table?

5. Where are elements with the highest boiling points located on the Periodic Table?

6. Explain any trends you noted in your answers to Questions 4 and 5.

7. Predict which element should have the lowest boiling point: selenium (Se), bromine (Br), or krypton (Kr). Explain how you decided.

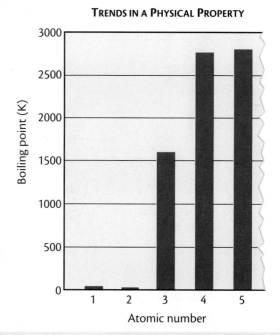

TRENDS IN A PHYSICAL PROPERTY

Figure 8 *Sample bar graph for boiling point data.*

When elements are listed in order of increasing atomic numbers and grouped according to similar properties, they form seven horizontal rows, called **periods.** This periodic relationship among elements is summarized in the modern Periodic Table, which you can see on page 104. To become more familiar with the Periodic Table, locate the 20 elements you grouped earlier. How do their relative positions compare with those shown on your chart?

PERIODIC TABLE OF THE ELEMENTS

Key

1	Atomic number
Hydrogen	Name
H	Symbol
1.008	Atomic weight

Legend:
- ▢ Metals
- ▢ Metalloids *(handwritten: semi-metal)*
- ▢ Nonmetals

Period	Group 1 (1A)	2 (2A)	3 (3B)	4 (4B)	5 (5B)	6 (6B)	7 (7B)	8 (8B)	9 (8B)	10 (8B)	11 (1B)	12 (2B)	13 (3A)	14 (4A)	15 (5A)	16 (6A)	17 (7A)	18 (8A)
1	1 H 1.008																	2 He 4.003
2	3 Li 6.94	4 Be 9.01											5 B 10.81	6 C 12.01	7 N 14.01	8 O 16.00	9 F 19.00	10 Ne 20.18
3	11 Na 22.99	12 Mg 24.31											13 Al 26.98	14 Si 28.09	15 P 30.97	16 S 32.07	17 Cl 35.45	18 Ar 39.95
4	19 K 39.10	20 Ca 40.08	21 Sc 44.96	22 Ti 47.87	23 V 50.94	24 Cr 52.00	25 Mn 54.94	26 Fe 55.85	27 Co 58.93	28 Ni 58.69	29 Cu 63.55	30 Zn 65.39	31 Ga 69.72	32 Ge 72.61	33 As 74.92	34 Se 78.96	35 Br 79.90	36 Kr 83.80
5	37 Rb 85.47	38 Sr 87.62	39 Y 88.91	40 Zr 91.22	41 Nb 92.91	42 Mo 95.94	43 Tc (98)	44 Ru 101.07	45 Rh 102.91	46 Pd 106.42	47 Ag 107.87	48 Cd 112.41	49 In 114.82	50 Sn 118.71	51 Sb 121.76	52 Te 127.60	53 I 126.90	54 Xe 131.29
6	55 Cs 132.91	56 Ba 137.33	71 Lu 174.97	72 Hf 178.49	73 Ta 180.95	74 W 183.84	75 Re 186.21	76 Os 190.23	77 Ir 192.22	78 Pt 195.08	79 Au 196.97	80 Hg 200.59	81 Tl 204.38	82 Pb 207.2	83 Bi 208.98	84 Po (209)	85 At (210)	86 Rn (222)
7	87 Fr (223)	88 Ra (226)	103 Lr (262)	104 Rf (261)	105 Db (262)	106 Sg (263)	107 Bh (264)	108 Hs (265)	109 Mt (268)	110 Uun (269)	111 Uuu (272)	112 Uub (277)		114 Uuq (285)		116 Uuh (289)		118 Uuo (293)

Lanthanides

57 La 138.91	58 Ce 140.12	59 Pr 140.91	60 Nd 144.24	61 Pm (145)	62 Sm 150.36	63 Eu 151.96	64 Gd 157.25	65 Tb 158.93	66 Dy 162.50	67 Ho 164.93	68 Er 167.26	69 Tm 168.93	70 Yb 173.04

Actinides

89 Ac (227)	90 Th 232.04	91 Pa 231.04	92 U 238.03	93 Np (237)	94 Pu (244)	95 Am (243)	96 Cm (247)	97 Bk (247)	98 Cf (251)	99 Es (252)	100 Fm (257)	101 Md (258)	102 No (259)

(handwritten annotations: "# electron shells", "# protons")

Each vertical column in the Periodic Table contains elements with similar properties. Each column is called a **group** or **family** of elements. For example, the lithium (Li) family, also called the **alkali metal** family, consists of the six elements (starting with lithium) in the first column at the left of the table. Each element (E) in this family is a highly reactive metal that forms an ECl chloride and E_2O oxide. By contrast, the helium family, at the right of the table, consists of very unreactive, and even inert, elements. Of the helium family, also called the **noble gas** family, only xenon (Xe) and krypton (Kr) are known to form compounds under normal conditions.

The arrangement of elements in the Periodic Table provides an orderly summary of the key characteristics of each element. By knowing the major properties of a certain chemical family, some of the behavior of any element in that family can be predicted. This can be very helpful in evaluating elements for certain uses.

CD-ROM WWW. **Trends & Variation in the Periodic Table**

Like sodium chloride (NaCl), all chlorides and oxides of lithium-family elements are ionic compounds.

PREDICTING PROPERTIES   : Building Skills 3

Some of an element's properties can be estimated by averaging the respective properties of the elements located just above and just below it. This is how Mendeleev predicted the properties of several elements unknown in his time. He was so convinced of the existence of these elements that he left gaps in the periodic table for them, along with some of their predicted properties. When those elements were eventually discovered, they fit exactly as expected. Mendeleev's fame rests largely on the accuracy of these predictions.

For example, germanium (Ge) was not known when Mendeleev proposed his periodic table. However, in 1871 he predicted the existence of germanium, calling it ekasilicon. Given that the boiling points of silicon (Si) and tin (Sn) are 3267 °C and 2603 °C, respectively, the boiling point of germanium can be estimated.

Eka- means "standing next in order."

The three elements are in the same group in the periodic table. Germanium is below silicon and above tin. (You can verify this by locating these elements in the Periodic Table.) Calculating the average of the boiling points of silicon and tin gives

$$\frac{(3267\,°C) + (2603\,°C)}{2} = 2935\,°C$$

When germanium was discovered in 1886, its boiling point was found to be 2834 °C. The estimated boiling point of germanium, 2935 °C, is within 4% of its known boiling point. The periodic table helped guide Mendeleev (and now you) to a useful prediction.

Formulas for chemical compounds can also be predicted from relationships established in the Periodic Table. For example, carbon and oxygen form carbon dioxide (CO_2). What formula would be predicted for a compound of carbon and sulfur? The Periodic Table indicates that sulfur (S) and oxygen (O) are in the same family. Knowing that carbon and oxygen form CO_2, a logical—and correct—prediction would be CS_2 (carbon disulfide). Now it's your turn.

1. The element krypton (Kr) was not known in Mendeleev's time. Given that the boiling point of argon (Ar) is −186 °C and of xenon (Xe) −112 °C, estimate the boiling point of krypton.

2. a. Estimate the melting point of rubidium (Rb). The melting points of potassium (K) and cesium (Cs) are 337 K and 302 K, respectively.

   b. Would you expect the melting point of sodium (Na) to be higher or lower than that of rubidium (Rb)? Explain.

3. Mendeleev knew that silicon tetrachloride ($SiCl_4$) existed. Using his periodic table, he correctly predicted the existence of ekasilicon, an element just below silicon in the Periodic Table. Predict the formula for the compound formed by Mendeleev's ekasilicon and chlorine.

4. Here are formulas for several known compounds: NaI, $MgCl_2$, CaO, Al_2O_3, and CCl_4. Using that information, predict the formula for a compound formed from

   a. C and F.   d. Ca and Br.
   b. Al and S.   e. Sr and O.
   c. K and Cl.

A.7 WHAT DETERMINES PROPERTIES?

Properties of Metals

What is responsible for differences in the number of chlorine atoms that react with a given atom—or in other properties that vary from element to element? Recall that atoms of different elements have different numbers of protons (atomic numbers). Therefore, atoms of different elements also have different numbers of electrons. Many properties of elements are determined largely by the number of electrons in their atoms and how these electrons are arranged.

A major difference between atoms of metals and nonmetals is that metal atoms lose some of their electrons much more easily than nonmetal atoms do. Under suitable conditions, one or more outer electrons of metal atoms may transfer to other atoms or ions. This is why metallic elements tend to form positive ions (cations).

Some physical properties of metals depend on attractions among their atoms. For example, stronger attractions among atoms of a metal result in higher melting points. The melting point of magnesium is 651 °C, whereas that of sodium is 98 °C. Thus attractions among the atoms in magnesium metal must be stronger than those in sodium metal.

Chemical and physical properties of nonmetals and compounds are also explained by the makeup of their atoms, ions, or molecules and by attractions among these particles. As you learned in Unit 1, the abnormally high melting and boiling points of water are due to the strong attractions among water molecules.

Understanding properties of atoms is the key to predicting and even manipulating the behavior of materials. Combined with a bit of imagination, this information allows chemists to find new uses for materials and to create new chemical compounds to meet specific needs.

You will learn more about electron arrangements in Unit 3.

Several thousand new compounds are synthesized each year.

A.8 IT'S ONLY MONEY

Based on what you have learned so far, you can start to make some decisions about your design for the new coin. A good first step is to specify some characteristics that are necessary or desirable in the material you will use. Apply your knowledge of existing coins, as well as what you have learned about properties of elements, to answer these questions.

1. What physical properties must material chosen for the new coin possess?

2. What other physical properties are desirable?

3. What chemical properties are required of the coin's material?

4. What other chemical properties are desirable?

5. Which would make the best primary material for the new coin: a metal, nonmetal, or metalloid? Explain.

6. What assumptions did you make in order to answer the preceding questions?

Save your answers to these questions—they will help guide your coin-design work later in this unit.

 Questions & Answers

The Elements (Values in parentheses are the mass numbers of the longest-lived isotopes.)

Element	Symbol	Atomic Number	Atomic Weight	Element	Symbol	Atomic Number	Atomic Weight
Actinium	Ac	89	(227)	Neodymium	Nd	60	144.24
Aluminum	Al	13	26.98	Neon	Ne	10	20.18
Americium	Am	95	(243)	Neptunium	Np	93	(237)
Antimony	Sb	51	121.76	Nickel	Ni	28	58.69
Argon	Ar	18	39.95	Niobium	Nb	41	92.91
Arsenic	As	33	74.92	Nitrogen	N	7	14.01
Astatine	At	85	(210)	Nobelium	No	102	(259)
Barium	Ba	56	137.33	Osmium	Os	76	190.23
Berkelium	Bk	97	(247)	Oxygen	O	8	16.00
Beryllium	Be	4	9.01	Palladium	Pd	46	106.42
Bismuth	Bi	83	208.98	Phosphorus	P	15	30.97
Bohrium	Bh	107	(264)	Platinum	Pt	78	195.08
Boron	B	5	10.81	Plutonium	Pu	94	(244)
Bromine	Br	35	79.90	Polonium	Po	84	(209)
Cadmium	Cd	48	112.41	Potassium	K	19	39.10
Calcium	Ca	20	40.08	Praseodymium	Pr	59	140.91
Californium	Cf	98	(251)	Promethium	Pm	61	(145)
Carbon	C	6	12.01	Protactinium	Pa	91	231.04
Cerium	Ce	58	140.12	Radium	Ra	88	(226)
Cesium	Cs	55	132.91	Radon	Rn	86	(222)
Chlorine	Cl	17	35.45	Rhenium	Re	75	186.21
Chromium	Cr	24	52.00	Rhodium	Rh	45	102.91
Cobalt	Co	27	58.93	Rubidium	Rb	37	85.47
Copper	Cu	29	63.55	Ruthenium	Ru	44	101.07
Curium	Cm	96	(247)	Rutherfordium	Rf	104	(261)
Dubnium	Db	105	(262)	Samarium	Sm	62	150.36
Dysprosium	Dy	66	162.50	Scandium	Sc	21	44.96
Einsteinium	Es	99	(252)	Seaborgium	Sg	106	(263)
Erbium	Er	68	167.26	Selenium	Se	34	78.96
Europium	Eu	63	151.96	Silicon	Si	14	28.09
Fermium	Fm	100	(257)	Silver	Ag	47	107.87
Fluorine	F	9	19.00	Sodium	Na	11	22.99
Francium	Fr	87	(223)	Strontium	Sr	38	87.62
Gadolinium	Gd	64	157.25	Sulfur	S	16	32.07
Gallium	Ga	31	69.72	Tantalum	Ta	73	180.95
Germanium	Ge	32	72.61	Technetium	Tc	43	(98)
Gold	Au	79	196.97	Tellurium	Te	52	127.60
Hafnium	Hf	72	178.49	Terbium	Tb	65	158.93
Hassium	Hs	108	(265)	Thallium	Tl	81	204.38
Helium	He	2	4.003	Thorium	Th	90	232.04
Holmium	Ho	67	164.93	Thulium	Tm	69	168.93
Hydrogen	H	1	1.008	Tin	Sn	50	118.71
Indium	In	49	114.82	Titanium	Ti	22	47.87
Iodine	I	53	126.90	Tungsten	W	74	183.84
Iridium	Ir	77	192.22	Ununnilium	Uun	110	(269)
Iron	Fe	26	55.85	Unununium	Uuu	111	(272)
Krypton	Kr	36	83.80	Ununbium	Uub	112	(277)
Lanthanum	La	57	138.91	Ununquadium	Uuq	114	(285)
Lawrencium	Lr	103	(262)	Ununhexium	Uuh	116	(289)
Lead	Pb	82	207.2	Ununoctium	Uuo	118	(293)
Lithium	Li	3	6.94	Uranium	U	92	238.03
Lutetium	Lu	71	174.97	Vanadium	V	23	50.94
Magnesium	Mg	12	24.31	Xenon	Xe	54	131.29
Manganese	Mn	25	54.94	Ytterbium	Yb	70	173.04
Meitnerium	Mt	109	(268)	Yttrium	Y	39	88.91
Mendelevium	Md	101	(258)	Zinc	Zn	30	65.39
Mercury	Hg	80	200.59	Zirconium	Zr	40	91.22
Molybdenum	Mo	42	95.94				

SECTION SUMMARY

Reviewing the Concepts

- The physical properties of a substance can be determined without altering the substance's chemical makeup; physical changes alter a substance's physical properties. Chemical properties describe a substance's tendency to react chemically; chemical changes transform the substance into one or more new substances.

1. Classify each as a chemical or physical property:

   a. Copper has a reddish brown color.
   b. Iron may rust when left outdoors.
   c. Carbon dioxide gas can extinguish a flame.
   d. Molasses pours more slowly than water.

2. Classify each as a chemical or physical change:

   a. A candle burns.

   b. A carbonated beverage fizzes when the container is opened.
   c. Hair curls as a result of a "perm."
   d. As shoes wear out, holes appear in the soles.

3. a. List the steps you would complete in making chocolate-chip cookies.
   b. Classify each step as involving either a chemical or a physical change.

- Elements can be classified as metals, nonmetals, or metalloids according to their physical and chemical properties.

4. Classify each property as characteristic of metals, nonmetals, or metalloids.

   a. Shiny in appearance
   b. Does not react with acids
   c. Shatters easily
   d. Dull in appearance but electrically conductive

5. Classify each element as a metal, nonmetal, or metalloid.

   a. tungsten
   b. antimony
   c. krypton
   d. sodium

6. What would you expect to happen if you hit a sample of each of these elements with a hammer?

   a. iodine
   b. zirconium
   c. phosphorus
   d. nickel

7. What properties make nonmetals unsuitable for electric wiring?

- Elements are arranged in rows (periods) on the Periodic Table. Elements with similar chemical properties are placed in the same vertical columns (groups or families).

8. Given the formulas AlN and $BeCl_2$, predict the formula for a compound containing

   a. Mg and F.
   b. Ga and P.

9. The melting points of oxygen and selenium are $-218\,°C$ and $221\,°C$, respectively. Estimate the melting point of sulfur.

10. Would you expect the boiling point of chlorine to be higher or lower than that of iodine? Explain.

11. For medical reasons, people with high blood pressure are advised to limit the amount of sodium ions in their diet. Normal table salt (NaCl) is sometimes replaced by a substitute called Lite Salt (potassium chloride) for seasoning.

   a. Write the formula for Lite Salt.
   b. Why are the properties of Lite Salt similar to those of sodium chloride?

♦ **The number of protons in an atom (the atomic number) of a given element distinguishes it from atoms of all other elements.**

12. Complete the table to the right for each electrically neutral atom.

13. A student is asked to explain the formation of a lead(II) ion (Pb^{2+}) from an electrically neutral lead atom (Pb). The student says that a lead atom must have gained two protons to make the ion. How would you correct this student's explanation?

Element symbol	Number of protons	Number of neutrons	Number of electrons
a. _____	6	6	6
b. _____	6	7	6
Ca	c. _____	21	d. _____
e. _____	f. _____	117	78
U	g. _____	146	h. _____

♦ **The mass of an atom depends largely on the number of protons and neutrons contained within its nucleus. Atoms containing the same number of protons but different numbers of neutrons are considered isotopes of that element.**

14. Supply the numbers of protons and neutrons for each of the isotopes in the table on the right.

15. A scientist announces the discovery of a new element. The only characteristic given in the report is the element's mass number of 266. Is this information sufficient, by itself, to justify the claim of the discovery of a new element? Explain.

Element symbol	Mass number	Number of protons	Number of neutrons
Mg	24	a. _____	b. _____
Mg	25	c. _____	d. _____
Mg	26	e. _____	f. _____

♦ **The properties of an element are determined largely by the number and arrangement of electrons in its atoms.**

16. Predict whether each element would be more likely to form an anion or a cation.

a. sodium
b. calcium
c. fluorine
d. oxygen
e. lithium
f. iodine

17. Noble gas elements rarely lose or gain electrons. What does this indicate about their chemical reactivity?

Connecting the Concepts

18. Which pair is more similar chemically? Defend your choice.

    a. copper metal and copper(II) ions
    or
    b. oxygen with mass number 16 and oxygen with mass number 18

19. Three kinds of observations that may indicate a chemical change are listed below. However, a physical change may also result in each observation. Describe a possible chemical cause *and* a possible physical cause for each observation.

    a. change in color
    b. change in temperature
    c. formation of a gas

20. The diameter of a magnesium ion (Mg^{2+}) is 156 pm (picometers, where 1 pm $=10^{-12}$ m); the diameter of a strontium ion (Sr^{2+}) is 254 pm. Estimate the diameter of a calcium ion (Ca^{2+}).

21. Identify the element in the Periodic Table described by each statement:

    a. This element is a member of a group of nonmetals. It forms anions with a −1 charge. It is in the same period as the metals used to form a penny.
    b. This element is a metalloid. It is in the same period as the elements found in table salt.

22. Mendeleev arranged elements in his periodic table in order of their atomic weights. In the modern Periodic Table, however, elements are arranged in order of their atomic numbers. Cite two examples from the Periodic Table for which these two schemes would produce a different ordering of adjacent elements.

Extending the Concepts

23. How is mercury different from other metallic elements? Using outside resources, describe some applications that take advantage of the unique properties of metallic mercury.

24. Depending on how it is heated and cooled, iron can either be hard and brittle or malleable. Explain how the same metal can have both characteristics.

25. Construct a graph of the price per gram of an element versus its atomic number for each of the first twenty elements. Can the current cost of those elements be regarded as a periodic property? Explain. (*Hint:* Use a chemical supply catalog or the Web to locate the current price of each element.)

CD-ROM
WWW.

26. Classify one or more pieces of jewelry you might possibly wear as being composed of metals, nonmetals, or metalloids.

EARTH'S MINERAL RESOURCES

SECTION B

Among Earth's resources, metals—and the minerals from which they are extracted—have had long-standing importance for and use by humans. Those uses have ranged from toolmaking, energy transmission, and construction to works of art, decoration, and coin making. In this section, you will explore the properties and uses of minerals and metals. Using copper as a case study, you will learn about Earth's mineral resources and how they are converted to pure metals.

B.1 SOURCES AND USES OF METALS

Earth's Mineral Resources

Human needs for resources—whether to create a new coin, manufacture new clothing, construct a space-vehicle launch rocket, or supply fertilizer for food crops—must all be met by chemical supplies currently present on Earth. These supplies of resources are often cataloged by where they are found. The table in Figure 9 indicates the composition of Earth.

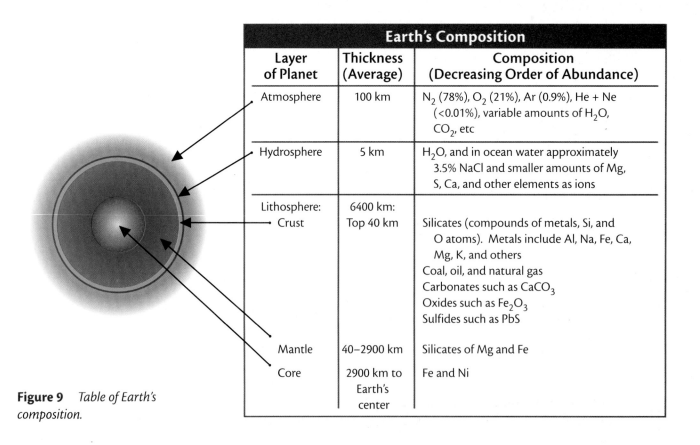

Earth's Composition		
Layer of Planet	Thickness (Average)	Composition (Decreasing Order of Abundance)
Atmosphere	100 km	N_2 (78%), O_2 (21%), Ar (0.9%), He + Ne (<0.01%), variable amounts of H_2O, CO_2, etc
Hydrosphere	5 km	H_2O, and in ocean water approximately 3.5% NaCl and smaller amounts of Mg, S, Ca, and other elements as ions
Lithosphere: Crust	6400 km: Top 40 km	Silicates (compounds of metals, Si, and O atoms). Metals include Al, Na, Fe, Ca, Mg, K, and others Coal, oil, and natural gas Carbonates such as $CaCO_3$ Oxides such as Fe_2O_3 Sulfides such as PbS
Mantle	40–2900 km	Silicates of Mg and Fe
Core	2900 km to Earth's center	Fe and Ni

Figure 9 *Table of Earth's composition.*

Earth's atmosphere, hydrosphere, and outer layer of the lithosphere (the solid part of Earth) supply all resources for all human activities. The atmosphere provides nitrogen, oxygen, neon, argon, and a few other gases. From the hydrosphere come water and some dissolved minerals. The lithosphere provides the greatest variety of chemical resources. For example, petroleum and metal-bearing ores are found there. An **ore** is a naturally occurring rock or mineral that can be mined and from which it is profitable to extract a metal or other material. An ore contains a mixture of components. Of these, the most important are **minerals**—solid compounds containing the element or group of elements of interest.

The deepest mines on Earth barely scratch the surface of its crust. If Earth were the size of an apple, all accessible resources of the lithosphere would be located within the apple's skin. From this thin band of soil and rock, we obtain the major raw materials needed to build homes, automobiles, appliances, computers, videotapes, compact discs, and sports equipment—in fact, all manufactured objects.

As you can see from the table in Figure 10 on page 114, many of Earth's important resources are not uniformly distributed. There is no connection between a nation's supply of these resources and either its land area or its population. Quite often a particular region serves as the predominant supplier of certain metals vitally important to industry. For example, Africa holds much of the world's known reserves of chromium (80%), cobalt (54%), and manganese (61%).

The development of the United States as a major industrial nation has been facilitated, in part, because of the quantity and diversity of chemical resources found here. Yet in recent years, the United States has imported increasing amounts of certain vital chemical resources. For example, about 75% of the nation's tin (Sn) is imported.

The greatest challenge regarding mineral resources is deciding on the wisest uses of the available supplies. Immediate issues include some that have technical and economic implications. For example, is it worthwhile to mine a particular metallic ore at a certain site? The answer depends on several factors:

- amount of useful ore at the site
- percent of metal in the ore
- type of mining and processing needed to extract the metal from its ore
- distance of the mine from metal refining facility and markets
- metal's supply-versus-demand status

Copper, one of the materials you might be thinking of using in your coin design, provides a case study of a vital chemical resource. You will first consider worldwide sources of copper and how these copper-bearing materials are converted to pure copper. Later, you will explore some possible replacements for this resource.

Copper is one of the most familiar and widely used metals in modern society. Among all the elements, it is second only to silver in electrical conductivity. This property and its relatively low cost, corrosion resistance, and ductility (ease of being drawn into thin wires), make copper the world's most common metal for electrical wiring. Copper is also used to produce

Worldwide Annual Production of Selected Metals

Metal	Country	Percent production	Actual production (1000 metric tons)	World total production (1000 metric tons)	Year
Aluminum	United States	17%	3713	22 100	1998
	Russia	14%	3005		
	Canada	11%	2374		
	China	10%	2100		
	Australia	7%	1627		
Copper[†]	United States	15%	1720	11 100	1997
	Chile	13%	1389		
	Japan	12%	1350		
	China	9%	963		
	Russia	5%	600		
Iron ore[○]	United States	22%	65 900	305 300	1997
	Brazil	12%	37 300		
	Russia	11%	34 000		
	Ukraine	10%	32 000		
	Canada	9%	27 300		
Lead[□]	United States	22%	1450	5880	1998
	United Kingdom	6%	350		
	Germany	6%	335		
	France	5%	306		
	Japan	5%	302		
Nickel[△]	Russia	23%	260	1120	1997
	Canada	17%	190		
	New Caledonia	12%	137		
	Australia	11%	124		
	Indonesia	6%	72		
Silver[△]	Mexico	16%	2.7	16.4	1998
	United States	13%	2.1		
	Peru	12%	1.9		
	Australia	9%	1.5		
	China	9%	1.4		
Tin[†]	China	29%	61	213	1997
	Indonesia	19%	40		
	Malaysia	17%	36		
	Brazil	9%	19		
	Bolivia	8%	16		
Zinc[†]	China	18%	1500	3890	1997
	Canada	9%	743		
	Japan	8%	653		
	Republic of Korea	5%	390		
	Spain	5%	370		

† = world smelter production   ○ = world pelletizing capacity   □ = world refinery production   △ = world mine production

Figure 10 *Production of selected metals worldwide.*

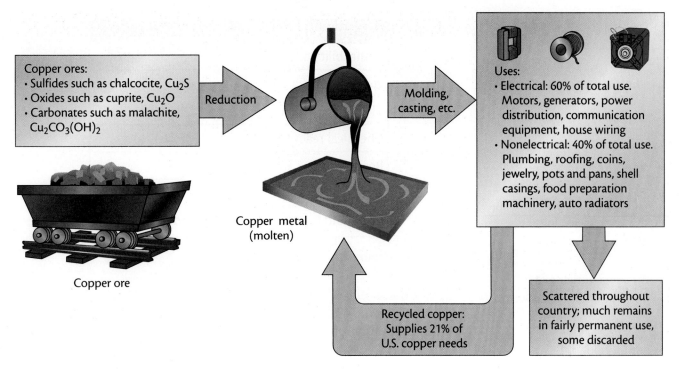

Copper ores:
- Sulfides such as chalcocite, Cu_2S
- Oxides such as cuprite, Cu_2O
- Carbonates such as malachite, $Cu_2CO_3(OH)_2$

Reduction

Copper ore

Copper metal (molten)

Molding, casting, etc.

Uses:
- Electrical: 60% of total use. Motors, generators, power distribution, communication equipment, house wiring
- Nonelectrical: 40% of total use. Plumbing, roofing, coins, jewelry, pots and pans, shell casings, food preparation machinery, auto radiators

Recycled copper: Supplies 21% of U.S. copper needs

Scattered throughout country; much remains in fairly permanent use, some discarded

Figure 11 *The copper cycle and copper uses.*

brass, bronze, and other alloys, a variety of important copper-based compounds, jewelry, and works of art.

The first copper ores mined contained from 35% to 88% copper. Although such ores are no longer available, ores less rich in copper can be used. In fact, it is now economically possible to mine ores containing less than 1% copper. Copper ore is chemically processed to produce metallic copper, which is then transformed into a variety of useful materials. Figure 11 summarizes the copper cycle from sources to common uses to waste products.

Earth's accessible deposits of this valuable resource are destined to be depleted. Will future developments increase or decrease the need for copper? What copper substitutes are available? The following activity will help you address these questions.

> Alloys are discussed in more detail in Section D.

PROPERTIES AND USES OF COPPER

Building Skills 4

Some of copper's properties are listed in the table in Figure 12 (page 116). Consider how these properties make copper suitable for uses depicted in Figure 11. For example, what properties make copper useful in electrical power generators? Copper's high electrical conductivity is certainly essential to this application. Copper's malleability and ductility are also important, making it possible to form copper wires and to wrap them in a generator. Corrosion resistance is also a benefit in such large, expensive equipment or in other applications.

Properties of Copper	
Malleability and ductility	High
Electrical conductivity	High
Thermal conductivity	High
Chemical reactivity	Relatively low
Resistance to corrosion	High
Useful alloys formed	Bronze, brass, etc.
Color and luster	Reddish, shiny

Figure 12 *Properties of copper.*

1. Consider the remaining uses of copper listed in Figure 11. For each use, identify those particular properties that make copper an appropriate choice.

2. a. How would increased recycling of scrap copper affect future availability of this metal?
   b. Is there a limit to the role copper recycling can play? Why?

3. For each use listed below, describe a technological change that could decrease the demand for copper:

   a. coins
   b. communications

   c. power generation
   d. indoor electrical wiring

B.2 CONVERTING COPPER   Laboratory Activity

Introduction

You have seen many chemical reactions in your lifetime. Some, such as a fireworks display, are memorable. Others, such as the slow process of rusting, are far less dramatic. Have you ever stopped to think about what happens to the atoms involved in those reactions? Are the materials that made up the fireworks still there after they are launched into the sky and ignited? What about the iron that turns into rust?

In this laboratory activity, you will work with a powdered sample of elemental copper, a metal you may be considering for your coin design. As you observe its chemical behavior, think about whether its properties make it a good candidate for this use.

Procedure

1. Prepare a data table to record the masses you will determine in Steps 2 and 9.

2. Measure and record the mass of a clean, empty crucible. Add approximately 1 g copper powder to the crucible. Record the mass of the crucible with copper powder in it within the nearest 0.1 g. Find

the actual mass of copper powder by subtracting the mass of the empty crucible from this value. Record the mass of copper powder.

3. Which properties of copper can be directly observed? Record your observations of the copper powder.

4. Set up the crucible, clay triangle, and burner as shown in Figure 13. The crucible lid should be slightly ajar.

5. Light the burner, and adjust it so that the flame tip just touches the bottom of the crucible.

6. Heat the crucible and its contents for two minutes. Remove the flame, and use a spatula to break up the solid in the crucible so that as much remaining copper metal is exposed as possible.

⚠ CAUTION: *Avoid touching the hot crucible.*

7. Continue heating for about 10 minutes more, removing the flame and breaking up the solid every two to three minutes.

8. When you have finished the heating, extinguish the burner flame and allow the crucible and its contents to cool to room temperature. Answer Questions 1 and 2 while you are waiting.

9. After the crucible and its contents have cooled, find their mass. Use this value and the mass of the empty crucible to calculate the mass of the contents. Record these values in your data table.

10. Transfer your product to a clean 100-mL beaker. Label and store the beaker and product as indicated by your teacher.

11. Put away the other materials. Wash your hands thoroughly before leaving the laboratory.

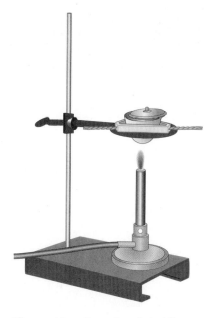

Figure 13 *Clay triangle holding a crucible over a burner.*

Questions

1. a. Were the changes you observed physical or chemical?
   b. How do you know?

2. a. Describe the changes you observed as you heated the copper.
   b. Did the copper atoms remain in the crucible? Explain.

3. a. What happened to the mass of the crucible contents after you heated the copper?
   b. Why do you think the mass changed?

B.3 METAL REACTIVITY

As you just observed, when copper metal is heated, it gradually reacts with oxygen in the air to produce a black substance. This is the equation for the reaction:

$$2\,Cu(s) \;+\; O_2(g) \;\longrightarrow\; 2\,CuO(s)$$

Copper   Oxygen   Copper(II) oxide

Although copper reacts to form copper(II) oxide when heated, at room temperature the metal remains relatively unreactive in air. You are probably familiar with this fact from observing that copper wire and the copper surface on pennies do not turn black under normal conditions.

There are two common compounds of copper and oxygen—CuO and Cu_2O. Because the name "copper oxide" could be applied to both, a Roman numeral is added to indicate copper's ionic charge. Copper(I) oxide is Cu_2O because it contains Cu^+ ions; copper(II) oxide is CuO because it contains Cu^{2+} ions.

Magnesium metal also reacts with oxygen gas. But unlike copper metal, magnesium heated in air ignites and produces a brief, blinding flash. See Figure 14. This is the equation for the reaction:

$$2\,Mg(s) \quad + \quad O_2(g) \quad \longrightarrow \quad 2\,MgO(s)$$
$$\text{Magnesium} \qquad\qquad \text{Oxygen} \qquad\qquad \text{Magnesium oxide}$$

By contrast, gold (Au) does not react with any components in air; in particular, it does not react with oxygen gas, even at elevated temperatures. This is one reason why gold metal is highly prized in long-lasting, decorative objects, such as jewelry. Gold-plated electrical contacts, such as those used for automobile air bags and audio cable connectors, are very dependable because nonconducting oxides do not form on the contact surfaces.

Observing how readily a certain metal reacts with oxygen provides information about the metal's reactivity. If elements are ranked in relative order of their chemical reactivity, the ranking is called an **activity series**. Based on what you have just learned about gold and magnesium and what you already know about copper, how would you rank the three metals in terms of their relative chemical reactivity?

B.4 RELATIVE REACTIVITIES OF METALS

Laboratory Activity

Introduction

In this activity, you will investigate the reactions of the metals copper, magnesium, and zinc with solutions that each contain a metal cation. The four solutions you will use are copper(II) nitrate, $Cu(NO_3)_2$ (containing Cu^{2+});

magnesium nitrate, $Mg(NO_3)_2$ (containing Mg^{2+}); zinc nitrate, $Zn(NO_3)_2$ (containing Zn^{2+}); and silver nitrate, $AgNO_3$ (containing Ag^+).

Procedure

1. Devise an orderly procedure that will allow you to observe the reaction (if any) between each metal and each of the four ionic solutions. You will conduct each reaction in a separate well of your wellplate, using five drops of 0.2 M solution and a small strip of metal. How many different combinations of metals and solutions will you need to observe? How will you arrange things so you can complete your observations efficiently yet remain certain which metal and which solution are in each well?

2. Prepare a data table to help you organize the observations and results of the procedure you devise.

3. Obtain 5-mm strips of each of the three metals to be tested. Clean the surface of each metal strip by rubbing it with sandpaper or emery paper. Record observations of each metal's appearance.

4. Complete your planned procedure, writing your observations in your data table. **CAUTION:** *Avoid letting the AgNO_3 solution come in contact with skin or clothing as it causes dark, non-washable stains.* If no reaction is observed, write NR in the table. Record the observed changes if a reaction occurs.

5. Dispose of your solid samples and wellplate solutions as directed by your teacher.

6. Wash your hands thoroughly before leaving the laboratory.

Questions

1. Which metal reacted with the most solutions?

2. Which metal reacted with the fewest solutions?

3. With which of the solutions (if any) would you expect silver metal to react, if it were available to be tested?

4. List the metals (including silver) in order, placing the most reactive metal first (the one reacting with the most solutions) and the least reactive metal last (the one reacting with the fewest solutions).

5. Refer to your "metal activity series" list in Question 4. Write a brief explanation of why the outside surface of a penny is made of copper instead of zinc.

6. a. Which of the four metals mentioned in this laboratory activity might be an even better choice than copper for the outside surface of a penny? Why?

   b. Why do you think that metal is not used for that purpose?

7. Given your new knowledge about the relative chemical activities of these four metals,

   a. which metal is most likely to be found in an uncombined, or "free," (metallic) state in nature?

   b. which metal is least likely to be found chemically uncombined with other elements?

Figure 15 *These stone, bronze, and iron tools represent three major ages of civilization.*

8. Reconsider your experimental design for this activity.

   a. Would it have been possible to eliminate one or more of the metal-solution combinations and still obtain all information needed to create chemical activity ratings for the metals?

   b. If so, which combination(s) could have been eliminated? Why?

B.5 METALS: PROPERTIES AND USES

Humans have been described as toolmakers. Readily available stone, wood, and natural fibers were the earliest materials used to make useful tools—hammers, chisels, knives, spears, and grinding devices. It was the discovery that fire could transform materials in certain rocks into strong, malleable metals, however, that triggered a dramatic leap in the growth of civilization.

Gold and silver, found as free elements rather than in chemical combination with other elements, were probably the first metals used by humans. These metals were formed into decorative objects and, later, into coins. Their relative unreactivity made them excellent materials for those uses.

It is estimated that copper has been used to make tools, weapons, utensils, and decorations for about 10 000 years. Bronze, an alloy of copper and tin, was developed about 3800 B.C. Thus humans moved from the Stone Age into the Bronze Age. See Figure 15.

Eventually early people developed iron metallurgy, the extraction of iron from its ores. This led to the start of the Iron Age, more than 3000 years ago. In time, as humans learned more about chemistry and fire, a variety of metallic ores were transformed into increasingly useful metals.

DISCOVERY OF METALS   ChemQuandary 1

Copper, gold, and silver are far from being the most abundant metals on Earth. Aluminum, iron, and calcium, for example, are all much more plentiful. Why, then, were copper, gold, and silver among the first metallic elements discovered?

You have explored some of the chemistry of metals and know, for example, that copper metal is more reactive than silver but less reactive than magnesium. A more complete activity series is given in the table in Figure 16. The table also includes brief descriptions of common methods for retrieving each metal from its ore.

In this list the most reactive metallic elements are at the top; less reactive elements are closer to the bottom. An activity list can be used to predict whether certain reactions can be expected. For example, you observed in the laboratory that zinc metal, which is more reactive than copper, reacted with copper ions in solution. However, zinc metal did not react with magnesium ions in solution. Why? Because zinc is less reactive than magnesium.

Metal Activity Series

Element	Metal Ion(s) Found in Minerals	Process Used to Obtain the Metal	State of Metal Obtained
Lithium	Li^+	Pass direct electric current through the molten mineral salt (electrometallurgy)	$Li(s)$
Potassium	K^+		$K(s)$
Calcium	Ca^{2+}		$Ca(s)$
Sodium	Na^+		$Na(s)$
Magnesium	Mg^{2+}		$Mg(s)$
Aluminum	Al^{3+}		$Al(s)$
Manganese	Mn^{2+}	Heat mineral with coke (C) or carbon monoxide (CO) (pyrometallurgy)	$Mn(s)$
Zinc	Zn^{2+}		$Zn(s)$
Chromium	Cr^{3+}, Cr^{2+}		$Cr(s)$
Iron	Fe^{3+}, Fe^{2+}		$Fe(s)$
Lead	Pb^{2+}	Heat (roast) mineral in air (pyrometallurgy) or find the element free (uncombined)	$Pb(s)$
Copper	Cu^{2+}, Cu^+		$Cu(s)$
Mercury	Hg^{2+}		$Hg(l)$
Silver	Ag^+		$Ag(s)$
Platinum	Pt^{2+}		$Pt(s)$
Gold	Au^{3+}, Au^+		$Au(s)$

Figure 16 *Metal activity series (in decreasing order of reactivity).*

[Mg(solid) causes Ag^+ (ions) to change into Ag(solid)]

In general, a more reactive metallic element (higher in the activity series) will cause ions of a less reactive metallic element (lower in the activity series) to change to their corresponding metal.

TRENDS IN METAL ACTIVITY — Building Skills 5

Use the table in Figure 16 and the Periodic Table (page 104) to answer the following questions.

1. a. What trend in metallic reactivity is found from left to right across a horizontal row (period) of the Periodic Table? (*Hint:* Compare the reactivities of sodium, magnesium, and aluminum.)
    b. In which part of the Periodic Table are the most-reactive metals found?
    c. Which part of the Periodic Table contains the least-reactive metals?

2. a. Will iron (Fe) metal react with a solution of lead(II) nitrate, $Pb(NO_3)_2$?
    b. Will platinum (Pt) metal react with a lead(II) nitrate solution?
    c. Explain your answers to Questions 2a and 2b.

3. Use specific examples from the activity series in your answers to these two questions:

    a. Are least-reactive metals also the cheapest metals?
    b. If not, what other factor(s) might influence the market value of a metal?

B.6 MINING AND REFINING

The process of converting a combined metal (usually a metal ion) in a mineral to a free metal involves a particular kind of chemical change. For example, the conversion of a copper(II) cation to an atom of copper metal requires two electrons.

Formation of Copper Metal

$$Cu^{2+} + 2\,e^- \longrightarrow Cu$$

Copper(II)   Copper
ion   metal

In general, to convert metal cations to neutral metal atoms, each cation must gain one or more electrons.

Chemists classify any chemical change in which a species gains one or more electrons as a **reduction.** Thus the conversion of copper(II) cations to copper metal is a reduction reaction. You can convince yourself of this fact by examining the equation above for that change. Chemists classify the reverse reaction, in which an ion or other species loses one or more electrons, as an **oxidation.** For example, under the right conditions copper atoms can be oxidized.

Formation of Copper(II) Ions

$$Cu \longrightarrow Cu^{2+} + 2\,e^-$$

Copper   Copper(II)
metal   ion

Historically, "oxidation" referred to the chemical combination of a substance with oxygen, as the term itself suggests. Chemists now know that in nearly all cases in which oxygen combines with another element or compound, oxygen partially or fully removes one or more electrons from the other species. By today's definition, any reactant—be it oxygen or not—that causes a species to lose one or more electrons is said to cause that species to be oxidized.

Whenever one species loses electrons, another species must simultaneously gain them. In other words, oxidation and reduction reactions never occur separately. Oxidation and reduction occur together in what chemists call **oxidation-reduction reactions** or, to use a common chemical nickname, **redox reactions.**

You have already observed redox reactions in the laboratory. In Laboratory Activity B.4 (pages 118–120), copper metal reacted with silver ions. Here is the oxidation-reduction reaction you observed:

$$Cu(s) + 2\,Ag^+(aq) \longrightarrow Cu^{2+}(aq) + 2\,Ag(s)$$

Copper   Silver   Copper(II)   Silver
metal   ion   ion   metal

> One way to remember this is **OIL RIG:** **O**xidation **I**s **L**oss of electrons, **R**eduction **I**s **G**ain of electrons.

A Copper Redox Reaction

Each metallic copper atom (Cu) was oxidized (converted to a Cu^{2+} ion by losing two electrons) and each silver ion (Ag^+ from $AgNO_3$ solution) was reduced (converted to an Ag atom by gaining one electron).

In the same activity you found that copper ions could be recovered from solution as copper metal by allowing the copper ions to react with magnesium metal, an element more active than copper. Magnesium atoms were oxidized; copper ions were reduced. Do you see why?

Note that the total electrical charge on both sides of this equation is the same. Electrical charges—as well as atoms—must balance in a correctly written chemical equation.

$$Cu^{2+}(aq) \; + \; Mg(s) \; \longrightarrow \; Cu(s) \; + \; Mg^{2+}(aq)$$

| Copper(II) ion | Magnesium metal | Copper metal | Magnesium ion |

In some circumstances this reaction might be a useful way to obtain copper metal. However, as is often the case, the desired copper metal is gained at the expense of "using up" another highly desirable material—in this case, magnesium metal.

How do redox reactions occur? Many metallic elements are found in minerals in the form of cations because they combine readily with other elements to form ionic compounds. Obtaining a metal from its mineral requires energy and a source of electrons. A reacting chemical species that serves as the source of electrons is known as a **reducing agent**.

Look again at Figure 16 (page 121). The table highlights several techniques that are used to reduce metal cations—or, in other words, to supply one or more electrons to each cation. The specific technique chosen depends on the metal's reactivity and the availability of inexpensive reducing agents and energy sources.

Two major approaches summarized in the table are **electrometallurgy** and **pyrometallurgy.** As the table suggests, electrometallurgy involves using an electric current to supply electrons to metal ions, thus reducing them. This process is used when no adequate chemical reducing agents are available or when very high-purity metal is sought. Pyrometallurgy—the most important and oldest ore-processing method—involves the treatment of metals and their ores by heat, as in a blast furnace. Carbon (coke) and carbon monoxide are common reducing agents in pyrometallurgy. A more active metal can be used if neither of these will do the job.

A third approach to obtaining metals from their ions is the process called **hydrometallurgy**—the treatment of ores and other metal-containing materials by reactants in water solution. You used such a procedure when you investigated the reactivity of different metals in Laboratory Activity B.4. Hydrometallurgy is used to recover silver and gold from old mine tailings (the mined rock left after most of the sought mineral is removed) by a process known as leaching. As supplies of higher-grade ores become scarcer, it will become economically feasible to use hydrometallurgy and other "wet processes" on metal-bearing minerals that dissolve in water.

ELECTRONS AND REDOX PROCESSES

The processes of oxidation (loss of one or more electrons) and reduction (gain of one or more electrons) can be clarified by visual representations of the events. To develop such representations, you will consider atoms of each of the metals you investigated in Laboratory Activity B.4.

First, however, a review of some key details about the composition of an atom is in order. Magnesium (Mg), an active metal, formed Mg^{2+} ions in several of the reactions. The atomic number of Mg is 12, indicating that an electrically neutral atom of magnesium contains 12 protons and 12 electrons. (Do you recall why those numbers must be equal for a neutral atom?)

If magnesium forms a Mg^{2+} ion, two negatively charged electrons must be removed from each magnesium atom. The bookkeeping involved in this change can be summarized this way:

Mg	$\longrightarrow$	Mg^{2+}	+	$2\,e^-$
12 protons (+)		12 protons (+)		
12 electrons (−)		10 electrons (−)		2 electrons (−)
Net charge: 0		Net charge: 2+		Net charge: 2−

To build a useful picture of this process in your mind, it is necessary to keep track of only the two electrons that each magnesium atom releases, rather than monitoring all 12 of the atom's available electrons. (In fact, in normal chemical reactions, a magnesium atom is not observed to release any of its other 10 electrons.)

Thus, for bookkeeping purposes, an atom of Mg will be depicted this way:

Mg: the symbol for the element with two dots attached.

Each dot represents one readily removable electron. The symbol Mg represents the remaining parts of a magnesium atom, including its other ten electrons. The resulting expression for Mg is called an **electron-dot structure,** or just a **dot structure.** The equation for the oxidation of Mg can be represented in electron-dot terms this way:

$$\text{Mg:} \longrightarrow Mg^{2+} + 2\,e^-$$

1. Construct a similar electron-dot expression for the change that occurred in Laboratory Activity B.4 when each of these events took place:

   a. An atom of zinc, Zn, was converted to a Zn^{2+} ion. (*Hint:* Zn has two readily removable electrons.)

   b. A silver ion, Ag^+, was converted to a metallic silver atom, Ag(s).

2. Apply the definitions of oxidation and reduction to your two equations in Question 1, and label each reaction appropriately.

Now consider one of the complete reactions that you observed in Laboratory Activity B.4. When you immersed a sample of copper metal, Cu, in silver nitrate solution, $AgNO_3$, a blue solution containing Cu^{2+} formed, as well as crystals of solid Ag. This is the reaction that occurred:

$$Cu(s) \; + \; 2\,Ag^+(aq) \; \longrightarrow \; Cu^{2+}(aq) \; + \; 2\,Ag(s)$$

Using dot structures, the reaction can be represented this way:

$$Cu\!: \; + \; Ag^+ \; + \; Ag^+ \; \longrightarrow \; Cu^{2+} \; + \; Ag\cdot \; + \; Ag\cdot$$

3. a. Which reactant (Cu or Ag^+) is oxidized?
    b. Which is reduced?

4. Why are two Ag^+ ions needed for each Cu(s) atom that reacts?

Each copper atom involved in this reaction loses two electrons. Thus copper atoms must be oxidized in the change. It is clear from the dot structures that those two electrons lost by copper are gained by two Ag^+ ions. So Ag^+ is the agent that caused the removal of electrons from Cu (resulting in the oxidation of Cu). The species involved in removing electrons from the reactant that is oxidized is called the oxidizing agent—in this case, Ag^+ ions.

5. a. Given that definition and explanation, what must be the reducing agent in the reaction between Cu(s) and Ag^+ ions?
    b. How would you define a reducing agent?

Now consider another reaction you observed in Laboratory Activity B.4:

$$Zn(s) \; + \; Cu^{2+}(aq) \; \longrightarrow \; Zn^{2+}(aq) \; + \; Cu(s)$$

6. Draw an electron-dot representation of this reaction.

7. a. Which reactant is oxidized?
    b. Which is reduced?

8. Identify the oxidizing agent and the reducing agent in this reaction.

9. Consider both of the oxidation-reduction reactions you analyzed in this exercise. What general features of an oxidation-reduction reaction would allow you to answer Questions 7 and 8 *without* drawing electron-dot representations?

10. Test your answer to Question 9 by considering a new oxidation-reduction reaction. Answer Questions 7 and 8 for this system:

$$Zn^{2+}(aq) \; + \; Mg(s) \; \longrightarrow \; Zn(s) \; + \; Mg^{2+}(aq)$$

SECTION SUMMARY

Reviewing the Concepts

♦ **The resources for all human activities must be obtained from Earth's atmosphere, hydrosphere, and outer layer of its lithosphere. These resources are not uniformly distributed.**

1. a. List and briefly describe the three major "parts" of Earth.
   b. Which part serves as the main storehouse of chemical resources used in manufacturing consumer products?

2. List two resources typically found in each of the three major parts of Earth.

3. According to information in Figure 10 on page 114, which of these four countries—the United States, Australia, China, or Brazil—produces the largest mass of these eight resources?

4. Is there a connection between the distribution of mineral resources and the wealth of a nation? Explain.

♦ **The feasible mining and extraction of a mineral resource depends, in part, on the amount of the resource available and the total cost of processing.**

5. What factors determine the feasibility of mining a particular metallic ore at a certain site?

6. A nineteenth-century gold mine, inactive for a hundred years, has recently reopened for further mining. What factors may have influenced the decision to reopen the mine?

7. What is meant by referring to the amount of "useful ore" at a site?

♦ **The ease with which a particular metal may be processed and preserved depends on its chemical reactivity. Active metals are more difficult to process than less-active metals and tend to corrode more quickly.**

8. Why are active metals more difficult to process and refine?

9. Based on your results from Laboratory Activity B.4, which metals involved in this activity would be the easiest to process? Why?

10. Why do most metals exist in nature as minerals rather than as pure metallic elements?

11. Consider these two equations. Which represents a reaction that is more likely to occur? Why?
    a. $Zn^{2+}(aq) + 2\,Ag(s) \longrightarrow Zn(s) + 2\,Ag^{+}(aq)$
    b. $2\,Ag^{+}(aq) + Zn(s) \longrightarrow 2\,Ag(s) + Zn^{2+}(aq)$

12. a. Why would it be a poor idea to stir a solution of lead(II) nitrate with an iron spoon?
    b. Write a chemical equation to support your answer.

♦ **The processes of oxidation (the loss of electrons) and reduction (the gain of electrons) occur together, resulting in oxidation-reduction (redox) reactions.**

13. Write an equation for each of these processes.
    a. the reduction of gold(III) ions
    b. the oxidation of elemental vanadium to vanadium(II) ions
    c. the oxidation of magnesium metal

14. Identify each equation as representing either an oxidation or a reduction reaction.
    a. $Fe^{2+} + 2\,e^{-} \rightarrow Fe$
    b. $Cr \rightarrow Cr^{3+} + 3\,e^{-}$
    c. $Al^{3+} + 3\,e^{-} \rightarrow Al$

15. Consider the following equation:
    $$Zn(s) + Ni^{2+}(aq) \longrightarrow Zn^{2+}(aq) + Ni(s)$$
    a. Which reactant has been oxidized? Explain your choice.
    b. Which reactant has been reduced? Explain your choice.
    c. What is the reducing agent in this reaction?

◆ **Metal cations can be converted to metal atoms by electrometallurgy, pyrometallurgy, or hydrometallurgy.**

16. Explain how each process converts metal cations to metal atoms.

    a. electrometallurgy

    b. pyrometallurgy

    c. hydrometallurgy

17. What processes could you use to obtain these elements from their ores?

    a. magnesium

    b. lead

Connecting the Concepts

18. How can a less active metal be used to prevent the corrosion of a more active metal?

19. Large gold nuggets with masses of 45 kg (100 pounds) or more have been discovered. What conditions might have allowed such large pieces of elemental gold to exist?

20. There are thousands of tons of gold in sea water. Explain why it is unlikely that ocean water will ever be "mined" for gold.

21. In 1982, when the penny was converted from pure copper to copper and zinc, the outside surface was still coated with copper. List three reasons why this coating was used.

22. At one time, food cans were made with tin and soldered with lead. What kinds of health hazards were posed by this arrangement?

23. Is there any connection between the process used to reduce a metal cation and the position of that element on the Periodic Table?

Extending the Concepts

24. Although aluminum is a more reactive metal than iron, it is often used for outdoor products. Investigate how this is possible.

25. The uneven distribution of mineral resources sometimes affects relations between nations. Identify and describe one historical or current example of this fact.

26. What conclusions about materials can be drawn from a study of the substances used for currency in ancient civilizations? Explain your ideas by giving examples.

27. History documents that copper has been used by humans for 10 000 years, whereas aluminum has been used for only about 100 years. Suggest and explain some reasons for this difference.

28. The reactive metal aluminum is often used in containers for acidic beverages. Investigate and describe the technology that makes this possible.

29. What is a patina? Explain its value both aesthetically and chemically.

CONSERVATION

Conservation

In some chemical reactions, matter seems to be created. In others, matter seems to disappear. Neither actually occurs. In this section, you will learn what happens to atoms in chemical reactions and how the atoms can be tracked through chemical equations. This information will help you to consider the fate of Earth's resources as well as the materials and products developed from them.

C.1 KEEPING TRACK OF ATOMS

Many things people use daily seem to disappear. For example, fuel is depleted as an airplane flies from one destination to another. The ice cream you eat seemingly vanishes. Steel in automobile bodies rusts away. Although the original forms of such materials disappear when they are used, the atoms composing them remain.

Think for a moment about what happens to molecules of gasoline as they burn in an automobile engine. The carbon and hydrogen atoms that make up these molecules react with oxygen atoms in the air to form carbon monoxide (CO), carbon dioxide (CO_2), and water (H_2O). These products are released as exhaust and disperse in the atmosphere. Thus the atoms of carbon, hydrogen, and oxygen originally present in the gasoline and air have not been destroyed; rather, they have been rearranged into new molecules.

In short, "using things up" means changing materials chemically, not destroying them. The law of conservation of matter, like all scientific laws, summarizes what has been learned by careful observation of nature: In a chemical reaction, matter is neither created nor destroyed. Molecules can be converted and decomposed by chemical processes, but atoms are forever. In a chemical reaction, matter—at the level of individual atoms—is always fully accounted for.

Because chemical reactions cannot create or destroy atoms, chemical equations representing such reactions must always be balanced. What does a "balanced equation" mean? Recall your introduction to chemical equations in Unit 1 (pages 29–31). Formulas for the reactants are placed on the left of the arrow; formulas for the products are placed on the right. In a balanced chemical equation, the number of atoms of each element is the same on the reactant and product sides.

Consider the burning of coal as an example. Coal is mostly carbon (C). If carbon burns completely, it combines with oxygen gas (O_2) to produce carbon dioxide (CO_2). Here is a representation of the atoms and molecules involved in this reaction:

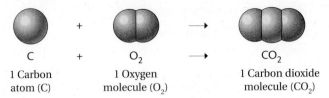

Note that the numbers of carbon and oxygen atoms on the reactant side equal the respective numbers of carbon and oxygen atoms on the product side. This indicates that the equation is balanced.

The representation of the coal-burning reaction shows that one carbon atom reacts with one oxygen molecule to form one carbon dioxide molecule. Written with chemical formulas, the equation becomes

$$C(s) \quad + \quad O_2(g) \quad \longrightarrow \quad CO_2(g)$$

Carbon and Oxygen React to Carbon
(in coal) gas produce dioxide gas

In writing chemical formulas for substances, the symbols for solid (s), liquid (l), and gas (g) are sometimes added. These symbols indicate the physical states of each substance under the conditions of the reaction.

The reaction in which you heated copper powder in air provides another example. Copper metal (Cu) reacts with oxygen gas (O_2) to form copper(II) oxide (CuO). Look at the representations below.

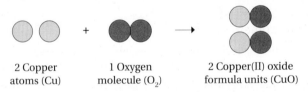

2 Copper   1 Oxygen   2 Copper(II) oxide
atoms (Cu)   molecule (O_2)   formula units (CuO)

Interpreting a Balanced Chemical Equation

Again, note that the numbers of copper and oxygen atoms on the reactant side equal the respective numbers of copper and oxygen atoms on the product side. Written as a chemical equation, the reaction is

$$2\,Cu(s) \quad + \quad O_2(g) \quad \longrightarrow \quad 2\,CuO(s)$$

Copper   and Oxygen React to Copper(II)
metal   gas   produce   oxide

You may have noticed that numbers have been placed in front of the copper and copper(II) oxide formulas. These numbers are called **coefficients**. Coefficients indicate the relative number of units of each substance involved in the chemical reaction. Reading this equation from left to right, you would say, "Two copper atoms react with one oxygen molecule to produce two formula units of copper(II) oxide."

Why is the term "formula unit" used instead of "molecule"? Compounds of a metal and a nonmetal are ionic. (It might be helpful to review pages 32–34 of Unit 1.) These compounds are not found as individual molecules. Rather, they form large crystals made of ions. Chemists use the term **formula unit** when referring to the smallest unit of an ionic compound.

In the following activity, you will practice recognizing and interpreting chemical equations. Then you will learn how to write them yourself.

It is standard not to write a coefficient of "1."

Look at the Periodic Table to recall which elements are metals and which are nonmetals.

ACCOUNTING FOR ATOMS

For each chemical statement that follows,

    a. interpret the statement in words,

    b. draw a representation of the chemical statement (some structures are provided),

    c. complete an atom inventory of the reactants and products, and

    d. decide whether the expression—as written—is balanced.

To help guide your work, the first item is worked out.

1. The reaction between propane (C_3H_8) and oxygen gas (O_2) is a common source of heat for campers, recreational-vehicle users, and others using tanks of liquid propane fuel. A chemical statement showing the reactants and products is

$$C_3H_8(g) + O_2(g) \longrightarrow CO_2(g) + H_2O(g)$$

> This reaction produces heat, which can also be considered a product.

    a. Interpreting this statement in words: "Propane gas reacts with oxygen gas to produce carbon dioxide gas and water vapor."

    b. Using ⚪⚫⚫⚫⚪ to represent a propane molecule, the chemical statement can be represented as:

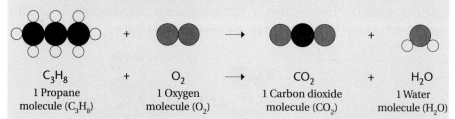

C_3H_8	+	O_2	→	CO_2	+	H_2O
1 Propane molecule (C_3H_8)		1 Oxygen molecule (O_2)		1 Carbon dioxide molecule (CO_2)		1 Water molecule (H_2O)

    c. Counting the atoms on each side of the equation gives this atom inventory:

Reactant side	Product side
3 carbon atoms	1 carbon atom
8 hydrogen atoms	2 hydrogen atoms
2 oxygen atoms	3 oxygen atoms

    d. The respective numbers of carbon, hydrogen, and oxygen atoms are different in reactants and products. Thus the original statement is not yet a properly written chemical equation. That is, it is not balanced.

2. Many people use natural gas to heat their homes. Natural gas contains methane (CH_4), which burns with oxygen (O_2) in air according to the equation

$$CH_4 + 2\,O_2 \longrightarrow CO_2 + 2\,H_2O$$

Use ⚪⚫⚪ to represent a methane molecule.

3. When an acid reacts with a metal, hydrogen gas and an ionic compound are often formed. An expression for hydrobromic acid (HBr)

reacting with magnesium metal to form hydrogen gas and magnesium bromide ($MgBr_2$) is:

$$HBr + Mg \longrightarrow H_2 + MgBr_2$$

Let ⬤⬤⬤ represent a formula unit of magnesium bromide ($MgBr_2$).

4. Hydrogen sulfide (H_2S) and metallic silver react in air to form silver sulfide (Ag_2S), commonly known as silver tarnish, and water:

$$4\,Ag + 4\,H_2S + O_2 \longrightarrow 2\,Ag_2S + 4\,H_2O$$

Let ⬤⬤⬤ represent a hydrogen sulfide molecule and ⬤⬤⬤ represent a formula unit of silver sulfide.

Try Questions 5 and 6 without drawing representations. (Why might this be a good decision?)

5. Wood or paper can burn in air to form carbon dioxide and water vapor. One component that burns is cellulose, represented by $C_6H_{10}O_5$.

$$C_6H_{10}O_5 + 6\,O_2 \longrightarrow 6\,CO_2 + 5\,H_2O$$

6. Nitroglycerin, $C_3H_5(NO_3)_3$, the active component of dynamite, decomposes explosively to form N_2, O_2, CO_2, and water.

$$2\,C_3H_5(NO_3)_3 \longrightarrow 3\,N_2 + O_2 + 6\,CO_2 + 5\,H_2O$$

> There are nine oxygen atoms in one $C_3H_5(NO_3)_3$ molecule. Can you see why?

C.2 NATURE'S CONSERVATION: BALANCED CHEMICAL EQUATIONS

The law of conservation of matter is based on the notion that Earth's basic "stuff"—its atoms—are indestructible. All changes observed in matter can be interpreted as rearrangements among atoms. Correctly written (balanced) chemical equations represent such changes. In the preceding activity, you practiced recognizing balanced chemical equations. Now you will learn how to write them.

As an example, consider the reaction of hydrogen gas with oxygen gas to produce gaseous water. First, write reactant formula(s) to the left of the arrow and product formula(s) to the right, keeping in mind that hydrogen and oxygen are diatomic (two-atom) molecules.

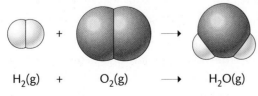

$$H_2(g) \quad + \quad O_2(g) \quad \longrightarrow \quad H_2O(g)$$

Check this expression by completing an atom inventory: Two hydrogen atoms appear on the left and two on the right. Hydrogen atoms are balanced. However, two oxygen atoms appear on the left and only one on the right. Because oxygen is not balanced, the expression requires additional work.

> Under certain conditions, this reaction produces a violent explosion. When controlled, the reaction powers some types of rockets. Used in fuel cells, it generates electricity.

Here is an *incorrect* way to complete the equation:

When balancing chemical equations, subscripts remain unchanged once the correct formulas have been written for the reactants and products. Coefficients must be adjusted instead.

$$H_2(g) \quad + \quad O_2(g) \quad \longrightarrow \quad H_2O_2(g) \quad \textbf{Incorrect!}$$

1 Hydrogen molecule   1 Oxygen molecule   1 Hydrogen peroxide molecule

Although this chemical statement satisfies atom-inventory standards (two hydrogen and two oxygen atoms on both sides), the expression is wrong. By changing the subscript of O from 1 to 2 in the product, the identity of the product has been changed from water (H_2O) to hydrogen peroxide (H_2O_2). Because hydrogen peroxide is not produced in this reaction, the expression is incorrect.

Additional hydrogen, oxygen, and water molecules must be added to the appropriate side of the equation to balance the numbers of oxygen and hydrogen atoms. Another oxygen atom is needed on the product side to bring the number of oxygen atoms on both sides to two. Therefore, a water molecule is added:

$$H_2 \quad + \quad O_2 \quad \longrightarrow \quad H_2O \quad + \quad H_2O$$

1 Hydrogen molecule   1 Oxygen molecule   2 Water molecules

Good! Two oxygen atoms appear on each side of the equation. Unfortunately, there are now two hydrogen atoms on the left and four on the right—hydrogen atoms are no longer balanced. How can two hydrogen atoms be added to the reactant side? You are correct if you said by adding one hydrogen molecule:

$$H_2 \quad + \quad H_2 \quad + \quad O_2 \quad \longrightarrow \quad H_2O \quad + \quad H_2O$$

2 Hydrogen molecules   1 Oxygen molecule   2 Water molecules

The atoms are balanced. Count them for yourself!

It is neither convenient nor efficient to draw representations for every chemical reaction. Thus this information is usually summarized in a chemical equation:

$$2\,H_2(g) + O_2(g) \longrightarrow 2\,H_2O(g)$$

The equation reads, "Two molecules of hydrogen gas react with one molecule of oxygen gas to produce two molecules of water vapor."

Here are some additional rules that may help you as you write chemical equations.

- If polyatomic ions, such as NO_3^- and CO_3^{2-}, appear as both reactants and products, treat them as units rather than balancing their atoms individually.

- If water is involved in the reaction, balance hydrogen and oxygen atoms last.

- Re-count all atoms after you think an equation is balanced—just to be sure!

WRITING CHEMICAL EQUATIONS

Building Skills 7

The reaction of methane gas (CH_4) with chlorine gas (Cl_2) occurs in sewage treatment plants and often in chlorinated water supplies. Common products are liquid chloroform ($CHCl_3$) and hydrogen chloride gas (HCl). A chemical statement describing this reaction follows below. As you can see from a quick glance, the equation is not balanced.

Chloroform is one of the THMs mentioned in Unit 1, page 74.

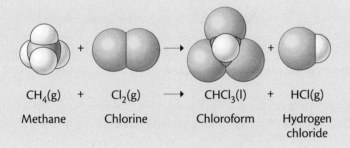

$CH_4(g)$	+	$Cl_2(g)$	→	$CHCl_3(l)$	+	HCl(g)
Methane		Chlorine		Chloroform		Hydrogen chloride

To complete the chemical equation, you can follow this line of reasoning: One carbon atom appears on each side of the arrow, so carbon atoms balance.

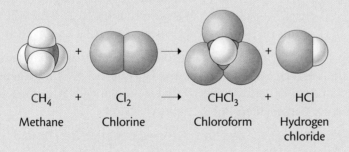

CH_4	+	Cl_2	→	$CHCl_3$	+	HCl
Methane		Chlorine		Chloroform		Hydrogen chloride

For convenience, the symbols (g) and (l) are removed. They will reappear in the final equation.

Four hydrogen atoms are on the left, but only two on the right (one in $CHCl_3$, a second in HCl). To increase the number of hydrogen atoms on the product side, the coefficient of HCl must be adjusted. Because two more hydrogens are needed on the right, the number of HCl molecules must be changed from 1 to 3. This gives four hydrogen atoms on the right:

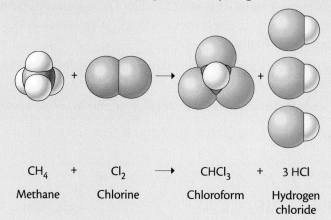

$$CH_4 \quad + \quad Cl_2 \quad \longrightarrow \quad CHCl_3 \quad + \quad 3\,HCl$$

| Methane | Chlorine | Chloroform | Hydrogen chloride |

Now both carbon and hydrogen atoms are balanced. What about chlorine? Two chlorine atoms appear on the left and six on the right side. These six chlorine atoms (three in $CHCl_3$, three in 3 HCl) must have come from three chlorine (Cl_2) molecules. Thus 3 must be the coefficient of Cl_2.

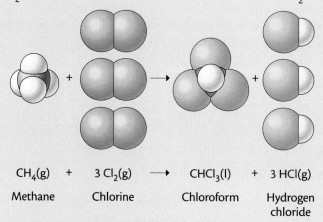

$$CH_4(g) \quad + \quad 3\,Cl_2(g) \quad \longrightarrow \quad CHCl_3(l) \quad + \quad 3\,HCl(g)$$

| Methane | Chlorine | Chloroform | Hydrogen chloride |

The chemical equation appears to be balanced. An atom inventory verifies that the equation is complete as written.

Reactant side	**Product side**
1 C atom	1 C atom
4 H atoms	4 H atoms (1 in $CHCl_3$, 3 in HCl)
6 Cl atoms (in 3 Cl_2)	6 Cl atoms (3 in $CHCl_3$, 3 in HCl)

Copy the following chemical expressions onto a separate sheet of paper, and balance each if needed. For Questions 1–4, draw a representation of your final equation to verify that it is balanced. Structures unfamiliar to you will be provided.

1. Two blast furnace reactions are used to obtain iron from its ore:

   a. $C(s) + O_2(g) \longrightarrow 2\,CO(g)$ b. $Fe_2O_3(s) + CO(g) \longrightarrow Fe(l) + 3\,CO_2(g)$

   Let ⬡ represent a formula unit of Fe_2O_3.

   Let ⬤ represent a molecule of CO_2.

2. The final step in the refining of a copper ore is:

$$CuO(s) + C(s) \longrightarrow Cu(s) + CO_2(g)$$

Let ◯⬤ represent a formula unit of CuO.

3. Ammonia (NH_3) in the soil reacts continuously with oxygen gas (O_2):

$$NH_3(g) + O_2(g) \longrightarrow NO_2(g) + H_2O(l)$$

4. Ozone (O_3) can decompose to form oxygen gas (O_2):

$$O_3(g) \longrightarrow O_2(g)$$

Let ⬤⬤⬤ represent an ozone molecule.

5. Copper metal reacts with silver nitrate solution to form copper(II) nitrate solution and silver metal:

$$Cu(s) + AgNO_3(aq) \longrightarrow Cu(NO_3)_2(aq) + Ag(s)$$

(*Hint:* Look at the formula for copper(II) nitrate, $Cu(NO_3)_2$. The subscript of two outside the parentheses indicates that this formula contains two nitrate (NO_3^-) anions. So one formula unit of $Cu(NO_3)_2$ contains one copper ion, two nitrogen atoms, and six oxygen atoms.)

6. Combustion of gasoline in an automobile engine can be represented by the burning of octane (C_8H_{18}):

$$C_8H_{18}(l) + O_2(g) \longrightarrow CO_2(g) + H_2O(g)$$

> Recall that in writing chemical formulas for substances, the symbols (s), (l), and (g), for solid, liquid, and gas respectively, are sometimes added. The symbol (aq) means that the substance is dissolved in water. It is an aqueous solution.

C.3 ATOM, MOLECULE, AND ION INVENTORY

In Question 2 in the previous activity, you obtained the chemical equation

$$2\,CuO(s) + C(s) \longrightarrow 2\,Cu(s) + CO_2(g)$$

One interpretation of this equation is, as you know: *Two formula units of copper(II) oxide and one atom of carbon react to produce two atoms of copper and one molecule of carbon dioxide.* Although correct, this interpretation involves such small quantities of material that a reaction on that scale would be completely unnoticed. Such information would not be very useful, for example, to a metal refiner interested in how much carbon is needed to react with a certain large-scale amount of copper(II) oxide.

Chemists have devised a counting unit called the **mole** (symbolized mol) that helps solve the refiner's problem. You are familiar with other counting units such as "pair" and "dozen." The mole can be regarded as the chemist's "dozen." A pair of water molecules is two water molecules. One dozen water molecules is 12 water molecules. One mole of water molecules is 602 000 000 000 000 000 000 000 water molecules. This number—the number of particles (or "things") in one mole—is more conveniently written as 6.02×10^{23}. Either way, this is a very large number!

> The number 6.02×10^{23} is called Avogadro's number.

Figure 17 *One mole each of copper, table salt, and water.*

Recall that mol is the symbol for the mole unit.

To help you get a better idea of the size of a mole, consider this: Imagine stringing a mole of paper clips (6.02×10^{23} paper clips) together and wrapping the string around the world. It would circle the world about 400 trillion (4×10^{14}) times! And even if you connected a million paper clips each second, it would take you 190 million centuries to finish stringing one mole of paper clips.

As large as one mole of molecules seems, however, drinking that amount of water would leave you quite thirsty on a hot day. One mole of water is less than one-tenth of a cup of water—only 18 g (or 18 mL) of water. But that is why the mole is so useful in chemistry. It represents a number of atoms, molecules, or formula units large enough to be conveniently weighed or measured in the laboratory. Furthermore, the atomic weights of elements can be used to find the mass of one mole of any substance, a value known as the **molar mass** of a substance. Figure 17 shows a mole of several familiar substances.

Specific examples will help you to better understand this notion. Suppose you need to find the molar masses of carbon (C) and of copper (Cu). In other words, you want to know the mass of one mole of carbon atoms (6.02×10^{23} atoms) and one mole of copper atoms (6.02×10^{23} atoms). Rather than counting that collection of atoms onto a laboratory balance (good luck!), you can quickly get the answers from atomic-weight data. The atomic weight of each element is found on the Periodic Table. Carbon's atomic weight is 12.01; copper's is 63.55. If the unit "grams" is added to these values, the result is their molar mass:

$$1 \text{ mol C} = 12.01 \text{ g C} \qquad 1 \text{ mol Cu} = 63.55 \text{ g Cu}$$

As you can now see, the mass (in grams) of one mole of an element's atoms equals the numerical value of the element's atomic weight. Any element's molar mass can be determined from the Periodic Table. The molar mass of a diatomic element is twice the mass given in the Periodic Table. Why?

The molar mass of a compound is the sum of the molar masses of its component atoms. For example, consider two compounds of interest to the copper-metal refiner—carbon dioxide (CO_2) and malachite ($Cu_2CO_3(OH)_2$).

One mole of CO_2 molecules contains one mole of C atoms and two moles of O atoms. Adding the molar mass of carbon and twice the molar mass of oxygen gives

$$1 \text{ mol C} \times \frac{12.01 \text{ g C}}{1 \text{ mol C}} = 12.01 \text{ g C}$$

$$2 \text{ mol O} \times \frac{16.00 \text{ g O}}{1 \text{ mol O}} = 32.00 \text{ g O}$$

Molar mass of CO_2 = (12.01 g + 32.00 g) = 44.01 g CO_2

The molar mass of malachite is found in a similar way. However, the total atoms of each element in more complex compounds must be carefully counted.

One mole of malachite ($Cu_2CO_3(OH)_2$) contains 2 mol Cu, 1 mol C, 5 mol O (can you see where all five oxygen atoms are found in the formula?), and 2 mol H.

$$2 \ \text{mol Cu} \times \frac{63.55 \ \text{g Cu}}{1 \ \text{mol Cu}} = 127.10 \ \text{g Cu}$$

$$1 \ \text{mol C} \times \frac{12.01 \ \text{g C}}{1 \ \text{mol C}} = 12.01 \ \text{g C}$$

$$5 \ \text{mol O} \times \frac{16.00 \ \text{g O}}{1 \ \text{mol O}} = 80.00 \ \text{g O}$$

$$2 \ \text{mol H} \times \frac{1.008 \ \text{g H}}{1 \ \text{mol H}} = 2.016 \ \text{g H}$$

$$\text{Molar mass of } Cu_2CO_3(OH)_2 = 221.13 \ \text{g } Cu_2CO_3(OH)_2$$

In summary, the molar mass of a compound is found by first multiplying the moles of each element in the formula by the molar mass of that element. Then all of the element masses are added together.

MOLAR MASSES   Building Skills 8

Find the molar mass of each substance:

1. The element nitrogen: N
2. Nitrogen gas: N_2
3. Sodium chloride (table salt): NaCl
4. Sucrose (table sugar): $C_{12}H_{22}O_{11}$
5. Chalcopyrite (a copper ore): $CuFeS_2$
6. Azurite (a copper ore): $Cu_3(CO_3)_2(OH)_2$ (*Hint:* Verify that this formula includes 8 moles of oxygen atoms.)

> The term "one mole of nitrogen" is confusing. It could refer to one mole of N atoms or a mole of N_2 molecules. Thus it is important to specify which particular substance is involved.

How are chemical equations and molar masses useful in answering questions about large-scale reactions? The coefficients in a chemical equation show both the relative numbers of individual molecules (or formula units) of reactants and products and the relative numbers of moles of these same substances. Consider again the metal-refining example. The mole "counting unit" makes it easy to find the mass of carbon dioxide released during refining.

$$2 \ CuO(s) \ + \ C(s) \ \longrightarrow \ 2 \ Cu(s) \ + \ CO_2(g)$$

| 2 Formula units CuO | 1 Atom C | 2 Atoms Cu | 1 Molecule CO_2 |

Also: 2 mol CuO   1 mol C   2 mol Cu   1 mol CO_2

Thus for every two moles of CuO that react, one mole of CO_2 is produced. The molar masses of all four substances can be used to interpret the equation in terms of the masses involved:

$$2 \ CuO(s) \ + \ C(s) \ \longrightarrow \ 2 \ Cu(s) \ + \ CO_2(g)$$

| 2 mol CuO | 1 mol C | 2 mol Cu | 1 mol CO_2 |
| 159.10 g CuO | 12.01 g C | 127.10 g Cu | 44.01 g CO_2 |

Compare the total mass of the products to the total mass of the reactants. How does this illustrate the law of conservation of matter?

> 2 mol CuO, containing 2 mol Cu atoms and 2 mol O atoms, has a mass of 159.10 g.

> $2 \ \text{mol Cu} \times \dfrac{63.55 \ \text{g Cu}}{1 \ \text{mol Cu}} = 127.10 \ g \ Cu$

Preserving the Past . . . for the Future

Mary Striegel is the Materials Research Program Manager at the National Center for Preservation Technology and Training (NCPTT) in Natchitoches, Louisiana. NCPTT is an effort by the National Park Service to advance the use of modern technologies in the practice of historic preservation in the fields of archaeology, architecture, landscape architecture, materials conservation, and history.

Mary supervises scientists in the Center's Materials Research Program, where field tests are used to determine the ways that air pollution and acid rain affect historic buildings, outdoor works of art, and various other materials. In laboratory work, the Center's scientists isolate different pollutants in small chambers in order to investigate the way materials and preservation treatments interact. In case studies, the scientists observe the effects of pollutants on artifacts that have been exposed in the environment.

> Scientist and art conservationists are combining their talents in an attempt to save these bronze treasures.

Preserving Art Treasures

Weathered to a powdery blue-green finish, bronze sculptures stand silently in city parks across the nation. Images of bygone heroes or memorials to noble historic struggles, the sculptures are gifts from previous generations. But bronze, an alloy made mostly of copper and tin, often corrodes in our modern industrial environment. Thus these links with the past are slowly being destroyed by the elements, pollution, acid rain, and changing temperatures. The problem is clearly evident on the pitted faces of the outdoor sculptures. Air pollution can ultimately lead to irregular patterns and decay of the metal—holes and streaks that are permanently etched into the piece.

But help may be on the way. Scientists and art conservationists are combining their talents in an attempt to save these bronze treasures.

The NCPTT has awarded a grant to researchers at North Dakota State University and the National Gallery of Art in Washington, D.C. The researchers' task is to find more effective methods of testing and developing a coating system that will help outdoor bronze sculpture better resist corrosion. The coating, however, must not yellow or change the sculpture's appearance. Five new coating systems are currently being tested both with accelerated corrosion test methods and under natural exposure to corrosive environments.

The two coating systems that are showing the most promise in the initial stages of testing are each made of acrylic-urethane and wax. The researchers will try to develop a variety of successful coatings because what works well in a tropical environment may not be the best choice in a northern climate. Likewise, the coating that is most effective in a city with heavy industry may not be ideal for a rural location.

"Not only will the development and testing of these coating systems affect sculpture in our country, but there is the possibility it will impact how we take care of sculptures worldwide," Striegel said.

Photograph courtesy of NCPTT

How Does Science Happen?

1. Scientific research may result in advances that improve our ability to deal with problems. Science is a cooperative venture: discoveries are often made by various groups and individuals working toward a common goal. How does the research described here demonstrate the cooperative nature of science?

2. How can a coating work to protect an exposed surface? Describe some of the chemistry behind the use of coatings in this kind of preservation effort.

3. Use the Web to find out about other projects of the National Center for Preservation Technology and Training (NCPTT).

C.4 COMPOSITION OF MATERIALS

One of the decisions you will make when designing your coin is whether to use only one material or a combination of materials. If your design uses more than one material, you will need to specify how much of each material will be present in the coin. The percent by mass of each material found in an item such as a coin is called its **percent composition**.

In Section A, you learned that the composition of the U.S. penny has changed several times. During 1943, it was made of zinc-coated steel. After this date and up until 1982, the penny was made mostly of copper. Since 1982, U.S. pennies have been made primarily of zinc. A post-1982 penny has a mass of 2.500 g and is composed of about 2.4375 g zinc and 0.0625 g copper. The percent composition of the penny can be found by dividing the mass of each constituent metal by the mass of the penny and multiplying by 100%:

$$\frac{2.4375 \text{ g zinc}}{2.500 \text{ g penny}} \times 100\% = 97.50\% \text{ zinc}$$

$$\frac{0.0625 \text{ g copper}}{2.500 \text{ g penny}} \times 100\% = 2.50\% \text{ copper}$$

The idea of percent composition also helps geologists describe how much metal or mineral is present in a particular ore. They can then evaluate whether the ore should be mined and also how it should be processed.

A compound's formula indicates the relative number of atoms of each element present in the substance. For example, one common commercial source of copper metal is the mineral chalcocite—copper(I) sulfide (Cu_2S). Its formula indicates that the mineral contains twice as many copper atoms as sulfur atoms. The formula also reveals how much copper can be extracted from a certain mass of the mineral—an important factor in copper mining and production.

How are the ideas of molar mass and percent composition useful in determining how much copper can be obtained from copper-containing minerals and ores? Some copper-containing minerals are listed in the table in Figure 18. The percent of copper in chalcocite can be found by applying what you know about molar masses.

The formula for chalcocite indicates that one mole of Cu_2S contains two moles of Cu, or 127.10 g Cu. The molar mass of Cu_2S is $(2 \times 63.55 \text{ g}) + 32.07 \text{ g} = 159.17 \text{ g}$. Therefore,

$$\% \text{ Cu} = \frac{\text{Mass of Cu}}{\text{Mass of Cu}_2\text{S}} \times 100 =$$

$$\frac{127.10 \text{ g Cu}}{159.17 \text{ g Cu}_2\text{S}} \times 100 = 79.85\% \text{ Cu}$$

Some Copper-Containing Minerals	
Common Name	**Formula**
Chalcocite	Cu_2S
Chalcopyrite	$CuFeS_2$
Malachite	$Cu_2CO_3(OH)_2$

Figure 18 *Names and formulas for three copper-containing minerals.*

A similar calculation indicates that Cu_2S contains 20.15% sulfur. The sum of percent copper and percent sulfur equals 100.00%. Why?

Knowing the percent composition of metal in a particular mineral is important in deciding whether a particular ore should be mined. What else needs to be considered? Suppose an ore is found to contain chalcocite, Cu_2S. Because nearly 80% of this mineral is composed of Cu (see calculation), it seems likely that this ore is worth mining for copper. However, another factor must be considered—the quantity of mineral contained in the ore. All other factors being equal, an ore that contains only 10% chalcocite would be less desirable as a copper source than one containing 50% chalcocite. Thus two factors must be taken into account to decide on the quality of a particular ore source: the percent mineral in the ore and the percent metal in the mineral.

Diagrams may be useful in understanding how these two percentage values relate to the total metal found in a particular ore. Consider an ore containing 10% chalcocite. Look at Figure 19. Suppose the rectangle in Figure 19a represents a piece of this ore. According to Figure 19b, 10% of the ore is composed of the mineral chalcocite. (One square is shaded with vertical stripes to show this.). You know that chalcocite itself is approximately 80% copper by mass. To represent this, 80% of the chalcocite square is shaded with horizontal stripes in Figure 19c. Now you can estimate visually how much copper is in this particular ore. How might similar diagrams represent an ore that is 50% chalcocite?

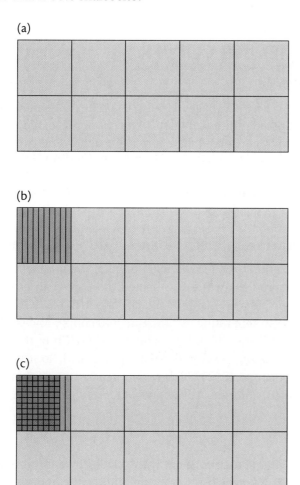

(a)

(b)

(c)

Figure 19 (a) *The rectangle represents the ore sample. Each square represents 10% of the sample.* (b) *One square is shaded with vertical stripes, representing an ore that is 10% chalcocite.* (c) *80% of the chalcocite is shaded with horizontal stripes. This represents the percent copper within chalcocite. Overall, only 8% of the ore is copper.*

PERCENT COMPOSITION

1. Use Figure 18 on page 140 to answer the following. In your calculations, assume that each mineral is present at the same concentration in an ore. Also assume that copper metal can be extracted from a given mass of any ore at the same cost.

   a. Calculate the percent copper in chalcopyrite.
   b. Calculate the percent copper in malachite.
   c. Which of the three minerals could be mined most profitably?

2. The chemical formula for the copper-containing mineral azurite is $Cu_3(CO_3)_2(OH)_2$. Complete an atom inventory for this compound. Use Building Skills 6 (page 130) as a model.

3. Suppose you have the choice either to mine ore that is 20% chalcocite (Cu_2S) or ore that is 30% chalcopyrite ($CuFeS_2$). Assuming all other factors are equal, which would you choose? Draw diagrams similar to those in Figure 19 on page 141 to support your answer.

4. Two common iron-containing minerals are hematite (Fe_2O_3) and magnetite (Fe_3O_4). If you had the same mass of each, which sample would contain the larger mass of iron? Support your answer with calculations.

> Recall that the percent copper in chalcocite was calculated on page 140.

C.5 RETRIEVING COPPER   Laboratory Activity

Introduction

In Laboratory Activity B.2 (pages 116–117), you heated metallic copper, producing a black powder that you know to be copper(II) oxide (CuO). Because atoms are always conserved in chemical reactions, the original copper atoms must still exist. In this laboratory activity, you will attempt to recover those atoms of metallic copper.

Procedure

Part I: Separating copper(II) oxide (CuO) from the sample

Most likely, in Laboratory Activity B.2, not all of the original copper powder reacted with oxygen gas when you heated the copper in air. Some copper metal is still likely mixed in with the black copper(II) oxide. The first steps in this activity involve separating this mixture into copper and copper(II) oxide. To do this, you will add dilute hydrochloric acid (HCl) to the black powder. Copper metal does not react with hydrochloric acid, so it will remain as a solid. The black copper(II) oxide, however, reacts with hydrochloric acid to produce copper(II) chloride ($CuCl_2$) and water, as shown in this equation:

$$CuO(s) \quad + \quad 2\,HCl(aq) \quad \longrightarrow \quad CuCl_2(aq) \quad + \quad H_2O(l)$$

Copper(II) oxide   Hydrochloric acid   Copper(II) chloride   Water

1. Obtain your beaker containing the black powder from Laboratory Activity B.2. Look closely at its contents. Is the material uniform throughout? If not, why not?

> Based on what you learned about mixtures on page 25, how would you classify the sample present in your beaker?

2. Add 50 mL of 1 M HCl to the beaker containing the copper oxide mixture. Record your observations of the solution. **CAUTION:** *Hydrochloric acid may damage your skin. If some HCl does spill on your skin, ask another student to notify your teacher immediately. Begin rinsing the affected area with tap water immediately.*

3. Gently heat the mixture to about 40 °C on a hot plate. Heat for 15 minutes, stirring every few minutes with a glass rod.

4. Remove the beaker from the heat source. Allow any unreacted copper metal remaining to settle to the bottom of the beaker. Then slowly decant the liquid into another 100-mL beaker.

5. Set aside the second beaker (containing the liquid) for Step 9.

6. Wash the solid copper remaining in the first beaker several times by swirling it gently with distilled water. Discard the liquid washings as instructed by your teacher.

7. Measure and record the mass of a piece of filter paper. Transfer the solid copper to the paper, and allow it to dry overnight.

8. When the sample and filter paper have dried, find the mass of the solid copper. This represents the portion of the original copper powder that failed to react to form CuO. Record this mass in your laboratory notebook.

Part II: Converting Copper(II) Chloride (CuCl₂) to Copper (Cu)

The final step is to convert the dissolved copper(II) chloride ($CuCl_2$) to copper metal. Perhaps you already have an idea how to do this. Recall the metal activity series that you devised in Laboratory Activity B.4 (pages 118–120). By adding solid metal samples to solutions containing other metal ions, you compared the chemical activity of each metal. Review your observations from that earlier laboratory activity. Which of the tested metals are more active than copper? Can you predict the result of placing one of those active metals in your copper(II) chloride solution? Check your prediction by performing these steps.

9. Obtain a watch glass that can completely cover the top of the beaker containing copper(II) chloride ($CuCl_2$) solution from Step 5. For each gram of copper powder that you started with in Laboratory Activity B.2, add about one gram of zinc metal to the $CuCl_2$ solution.

10. Immediately cover the beaker with the watch glass, and allow it to stand for several minutes. Record your observations.

11. After the reaction has subsided, remove the watch glass and gently dislodge solid copper that forms on the surfaces of the zinc pieces.

12. Continue to dislodge copper from the zinc until you are convinced that the zinc has stopped reacting with the solution. (How can you decide?) Then add 10 mL of 1 M HCl to the beaker and carefully remove any large pieces of solid zinc from the beaker with forceps. Replace the watch glass. Record your observations.

13. After a few minutes, carefully decant as much of the liquid as possible into another beaker.

⚠ Note warning about hydrochloric acid in Step 2.

14. Wash the solid copper several times with distilled water.

15. Transfer the copper to a preweighed piece of filter paper, and allow it to dry overnight.

16. When the sample and filter paper have dried, find the mass of copper metal. This represents the copper that you recovered from the copper(II) chloride solution.

17. Follow your teacher's instructions for disposing of waste materials.

18. Wash your hands thoroughly before leaving the laboratory.

Questions

1. In Laboratory Activity B.2 not all of the original copper powder reacted when you heated it in air.
   a. Why do you think the reaction was incomplete?
   b. How would you revise the procedure so that more copper(II) oxide could form?

2. a. In Laboratory Activity B.2 what mass of the original copper-powder sample reacted when you heated it? (*Hint:* Use the original mass of copper used in Laboratory Activity B.2 and the mass of copper residue found in Step 8 to calculate this.)
   b. What percent of the total copper-powder sample reacted?

3. In the reaction between copper(II) chloride ($CuCl_2$) solution and zinc metal, each Cu^{2+} ion gained two electrons to form an atom of copper metal. Each zinc metal atom lost two electrons to form a Zn^{2+} ion.
   a. Write a chemical equation that represents this process. (To review how, turn to pages 122–123.)
   b. Based on the equation you wrote in Question 3a, identify
      i. the species that was oxidized.
      ii. the species that was reduced.
      iii. the reducing agent.
      iv. the oxidizing agent.

4. Adding HCl to CuO resulted in the formation of a blue solution. This color is due to the presence of Cu^{2+}(aq) ions. Consult your observations in answering these questions:
   a. Describe what happened to the solution color after you added zinc in Step 9.
   b. What caused the changes you observed in the solution?
   c. How can the color of the solution be used to indicate when the zinc metal has removed the Cu^{2+} ions in solution?

5. To recover Cu metal from the $CuCl_2$ solution, you had to use other resources.
   a. What resources were "used up" in this process?
   b. Where did each of them go?

C.6 CONSERVATION IN THE COMMUNITY

In some ways Earth is like a spaceship. The resources "on board" are all that are available to the inhabitants of the ship. Some resources—such as fresh water, air, fertile soil, plants, and animals—can eventually be replenished by natural processes. These resources are called **renewable resources.** As long as natural cycles are not disturbed too much, supplies of renewable resources can be maintained indefinitely. Other materials—such as metals, natural gas, coal, and petroleum—are considered **nonrenewable resources** because they cannot be readily replenished. If atoms are always conserved, why do some people say that a resource may be "running out"? Can a resource actually "run out"?

The answer can be found by first remembering that atoms are conserved in chemical processes, but molecules might not be. For example, the current production of new petroleum molecules in nature is very much slower than the current rate at which petroleum molecules are being burned to produce carbon dioxide, water, and other molecules. Thus the total inventory of petroleum on Earth is declining.

A resource—particularly a metal—can be depleted in another way. As you learned in Section B, profitable mining depends on finding an ore with at least some minimum metal content. This minimum level depends on the metal and its ore: from as low as 1% for copper or 0.001% for gold to as high as 30% for aluminum.

Once ores with high metal content are depleted, lower-grade ores with less metal content are processed. Meanwhile, atoms of the metal, once concentrated in rich deposits within limited parts of the world, gradually become spread out (dispersed) over wider areas of Earth. Eventually, the mining and extraction of certain metals may become prohibitively expensive for general use. At that time, for practical purposes the supply of that resource can be considered depleted.

Can such depletion scenarios be avoided? Can Earth's mineral resources be conserved? One strategy for conservation is to slow down the rate at which the resources are used. Part of this strategy includes rethinking personal and societal habits and practices involving resource use. Such rethinking can involve decisions such as whether it is better to use paper or plastic bags in grocery stores, as well as whether, as is done in some parts of the world, it is even better to encourage customers to bring their own reusable bags.

Rethinking can take the form of re-examining old assumptions, identifying resource-saving strategies, and, perhaps, uncovering new solutions to old problems. Possibly the most important part of rethinking resource conservation and management is to consider the most direct option—source reduction. That simply means decreasing the amount of resources used. The fewer resources used, the more that remain available for future generations.

Another approach is to replace a resource by finding substitute materials with similar properties, preferably materials from renewable resources. In addition, some manufactured items can be refurbished or repaired for reuse rather than sent to a landfill. Common examples include used car parts and printer cartridges, both of which can often be reused after rebuilding or refilling. Finally, certain items can be recycled, or gathered for

> Resources such as petroleum are regarded as nonrenewable because they can be formed only over millions of years.

> Minimum profitability levels for mining depend not only on the abundance of the metal in the ore but also on the metal's chemical activity. Less active metals (such as gold) can more easily be released from their compounds than can more active metals (such as aluminum). Thus richer ores are needed to make the mining of active metals profitable.

reprocessing. Recycling allows the resources present in the items to be used again. Figure 20 illustrates the steps involved in recycling aluminum cans.

However, even the recycling process can create environmental problems. For example, aluminum scrap, excluding cans, may contain up to one percent magnesium. The magnesium must be removed before the aluminum is processed into other products. Conventional removal of magnesium requires the addition of chlorine gas, which requires special handling in the workplace and can contribute to air pollution. An alternative method uses ceramic oxides, a process that is safer to use and that eliminates chlorine emissions to the atmosphere.

Ultimately, some items will be discarded if they are no longer wanted or needed. Each person in this country throws away an average of about 2 kg (4 lb) of unwanted materials (or waste) daily. Some products, such as yesterday's newspaper, become waste after they fulfill their initial purpose. Others, such as telephones and computers, become waste when they are discarded for newer models. Combined, materials directly discarded by U.S. citizens would fill the New Orleans Superdome from floor to ceiling twice each day.

As you can see in Figure 21, the largest fraction of municipal waste in the United States is paper. While recycling reduces this component some-

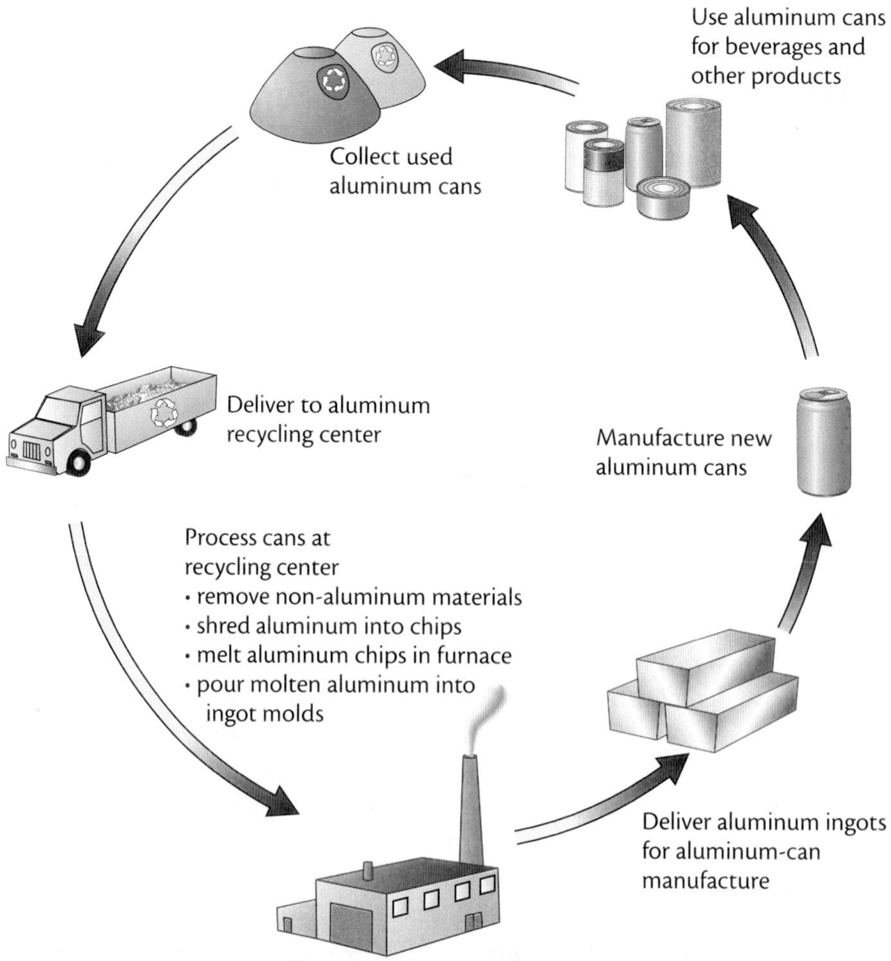

> In some nations most glass bottles are refilled, not discarded.

Figure 20 *Aluminum can recycling.*

Use aluminum cans for beverages and other products

Collect used aluminum cans

Deliver to aluminum recycling center

Process cans at recycling center
• remove non-aluminum materials
• shred aluminum into chips
• melt aluminum chips in furnace
• pour molten aluminum into ingot molds

Manufacture new aluminum cans

Deliver aluminum ingots for aluminum-can manufacture

what, the market for recycled paper is limited. Because this leaves a high proportion of combustibles in the waste stream, waste-to-energy plants have become an attractive option. More than 120 waste-to-energy plants currently operate in the United States, burning about 97 000 tons of solid waste per day. Each ton of garbage that serves as "fuel" in these plants produces about a third of the energy released by a similar quantity of coal.

Although waste-to-energy plants produce some fly ash and solid residues, such plants can allow the recycling of materials that otherwise would be disposed of as part of an unwanted product. In addition, waste-to-energy plants tend to increase recycling, both on-site and in the communities in which they are located.

Recycling, landfilling, and combustion for energy production are three options for the final step in the life cycle of a material—a topic you will now consider as you prepare to choose a material for your coin design.

WASTE GENERATED IN UNITED STATES IN 1997 BEFORE RECYCLING

Category	Weight (million tons)	Percent of Waste
Paper	83.8	38.6
Yard waste	27.7	12.8
Plastics	21.5	9.9
Metals	16.6	7.7
Wood	11.6	5.3
Food waste	21.9	10.1
Glass	12.0	5.5
Other	21.8	10.0

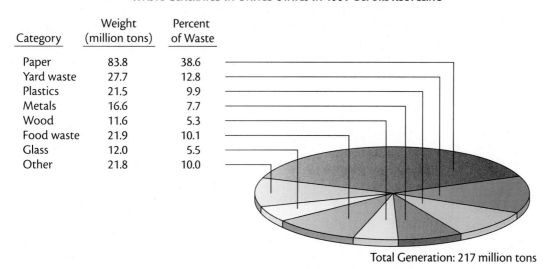

Total Generation: 217 million tons

WASTE GENERATED IN UNITED STATES IN 1997 AFTER RECYCLING

Category	Weight (million tons)	Percent of Waste
Paper	48.9	31.3
Yard waste	16.2	10.4
Plastics	20.4	13.0
Metals	10.1	6.4
Wood	11.0	7.0
Food waste	21.3	13.6
Glass	9.1	5.8
Other	19.3	12.3

Source: Characterization of MSW in the U.S. :
1998 Update, U.S. EPA, Washington, DC

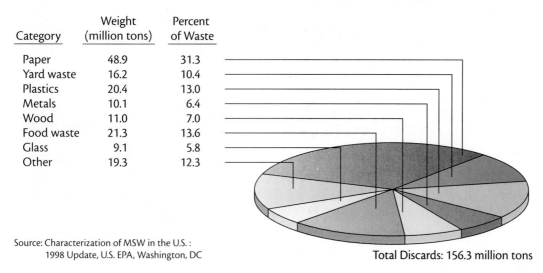

Total Discards: 156.3 million tons

Figure 21 *Composition of U.S. municipal waste stream before and after recycling.*

C.7 THE LIFE CYCLE OF A MATERIAL

In designing a new product for human use, proper evaluation must include consideration of the full life cycle of the materials involved. Such a life cycle has several distinct stages. Raw materials are first obtained and then refined and synthesized into the desired material. That material is then used to make the product designed for a particular use. When the product is no longer useful, the materials may be recovered and reused, or they may end up scattered in landfills. Figure 22 illustrates this general life cycle.

In every step of the cycle, energy and resources are used. Laboratory Activities B.2 (Converting Copper) and C.5 (Retrieving Copper) are good examples of this fact. Recall that heat energy and chemical resources (hydrochloric acid and zinc metal) were used first to convert the copper metal to other substances and then to recover the copper metal. Because energy use and resource use impact economics as well as the environment, each step in the life cycle of a material becomes a factor to consider when a new product is designed.

The next activity will allow you to model this process for a familiar material—copper.

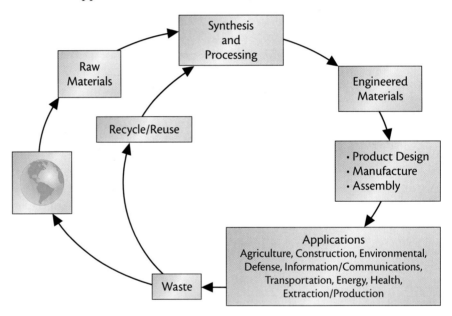

Figure 22 *Life cycle of materials.*

COPPER LIFE-CYCLE ANALYSIS   Building Skills 10

Imagine that you are involved in designing copper water pipes. Use Figure 11 (page 115) and Figure 22 to conduct a life-cycle analysis for the copper metal in your pipes. Consider how each step in the life cycle of copper will affect your final design by answering the following questions.

1. Which life-cycle steps consume significant quantities of energy?

2. Which steps (such as the reduction of a mineral in an ore to produce the metal) require the use of other materials?

3. How will you obtain the copper for your pipes?

4. Consider the transportation of materials in each step. How might this influence the design of your copper pipes?

5. Copper pipes may someday no longer be in use.

   a. What will happen to the copper in them?

   b. How can this issue be addressed when designing the pipes?

6. Consider your answers to each of the previous questions. How does each decision influence the cost of your copper pipes?

C.8 DESIGNING A COIN

Making Decisions

By now you have a good idea of what properties your half-dollar coin should possess. Perhaps you have even narrowed the range of materials still under consideration. To prepare for the final phase of your coin design, answer these questions.

1. Look at the list of required and desired physical and chemical properties you developed on page 107 of Section A. Revise and update your list using what you have learned since completing that activity.

2. Coins are traditionally made of metals. What alternative materials might be considered? What are the advantages or disadvantages of these alternatives?

3. Construct a table similar to the one shown below. Enter all necessary and desirable physical properties from your revised list in Question 1. Then list the materials you are considering for use in your coin. If you have not already done so, begin researching the properties of these materials. Indicate whether each material meets each criterion. For example, you might check the box if it does and place an "x" in the box if it does not. Construct a similar table for your list of necessary and desirable chemical properties.

		Materials				
		A	B	C	D	E
Necessary Properties						
Desirable Properties						

4. Consider the design of your coin.

    a. What images are you considering for the obverse ("heads") and reverse ("tails") sides of the coin?

    b. What anticounterfeiting features are appropriate to include in your design?

    c. How large should this coin be?

    d. What should be its mass?

5. The cost of materials should be considered in your design.

    a. What would happen if the materials in the coin were worth more than the face value of the coin (fifty cents)?

    b. What particular materials have you already ruled out because of this concern?

6. What other costs need to be considered as you design your coin?

Questions & Answers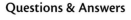

7. Consider the life cycle of your coin.

    a. How will the coin materials be obtained and processed?

    b. How long do you expect each coin to remain in circulation?

    c. What will happen to the materials when used coins are removed from circulation?

    d. How might impact on the environment be addressed at each stage of the coin's life cycle?

SECTION SUMMARY

Reviewing the Concepts

♦ **Matter and its constituent atoms are neither created nor destroyed in a chemical reaction.**

1. a. State the law of conservation of matter.
   b. How is a scientific law, such as the law of conservation of matter, different from a law created by the government?

2. Complete atom inventories to determine if each of the following equations is balanced:

   a. The preparation of tin(II) fluoride, an ingredient in some toothpastes (called "stannous fluoride" on some labels):

   $$Sn(s) + HF(aq) \longrightarrow SnF_2(aq) + H_2(g)$$
   Tin metal   Hydrofluoric   Tin(II)   Hydrogen
             acid   fluoride   gas

   b. The synthesis of carborundum for sandpaper:

   $$SiO_2(s) + C(s) \longrightarrow SiC(s) + CO(g)$$
   Silicon   Carbon   Silicon   Carbon
   dioxide   carbide   monoxide
   (sand)   (carborundum)

   c. The reaction of an antacid with stomach acid:

   $$Al(OH)_3(s) + 3\ HCl(aq) \longrightarrow AlCl_3(aq) + 3\ H_2O(l)$$
   Aluminum   Hydrochloric   Aluminum   Water
   hydroxide   acid   chloride

3. Why are the phrases "using up" and "throwing away" inaccurate from a chemical viewpoint?

♦ **One mole contains 6.02×10^{23} particles. The molar mass of a substance can be determined from the atomic weights of elements composing that substance.**

4. Find the molar mass of each substance.

   a. oxygen gas, O_2
   b. ozone, O_3
   c. caffeine, $C_8H_{10}N_4O_2$
   d. a typical antacid, $Mg(OH)_2$
   e. aspirin, $C_9H_8O_4$

5. Create an analogy that illustrates the mole concept. Refer to Figure 17 and the text in C.3, Atom, Molecule, and Ion Inventory (pages 135–137) for examples.

♦ **Coefficients in a chemical equation indicate the relative number of units of each reactant and product. Subscripts in a substance's formula may not be changed in order to balance an equation.**

6. Balance each of the following chemical expressions:

   a. Preparing phosphoric acid (used in making soft drinks, detergents, and other products) from calcium phosphate and sulfuric acid:

   $$\_\_Ca_3(PO_4)_2 + \_\_H_2SO_4 \longrightarrow \_\_H_3PO_4 + \_\_CaSO_4$$

   b. Preparing tungsten from one of its minerals:

   $$\_\_WO_3 + \_\_H_2 \longrightarrow \_\_W + \_\_H_2O$$

   c. Heating lead sulfide in air:

   $$\_\_PbS + \_\_O_2 \longrightarrow \_\_PbO + \_\_SO_2$$

   d. Burning gasoline:

   $$\_\_C_8H_{18} + \_\_O_2 \longrightarrow \_\_CO_2 + \_\_H_2O$$

   e. Rusting (oxidation) of iron metal:

   $$\_\_Fe + \_\_O_2 \longrightarrow \_\_Fe_2O_3$$

7. A student is asked to balance this chemical expression:

   $$Na_2SO_4 + KCl \longrightarrow NaCl + K_2SO_4$$

   The student decides to balance the expression this way:

   $$Na_2SO_4 + K_2Cl \longrightarrow Na_2Cl + K_2SO_4$$

   a. Complete an atom inventory on the student's answer. Are all atoms conserved?
   b. Did the student create a properly balanced chemical equation? Explain.
   c. If your answer to Question 7b is "no," write a correctly balanced equation.

- The percent composition of a substance can be calculated from the relative number of atoms of each element in the substance.

8. Find the percent metal (by mass) in each compound:

   a. Ag_2S
   b. Al_2O_3
   c. $CaCO_3$

9. A 50.0-g sample of ore contains 5.00 g lead(II) sulfate ($PbSO_4$).

   a. What is the percent lead metal (Pb) in $PbSO_4$?
   b. What is the percent $PbSO_4$ in the ore sample?
   c. What is the percent Pb in the total ore sample?

- Resources are either renewable or nonrenewable. Resources can be conserved by recycling, controlling rate of use, or replacing with substitute resources.

10. What is the difference between reusing and recycling? Give two examples of each, other than those presented in the textbook.

11. In addition to those found in this textbook, list four examples of

    a. renewable resources.
    b. nonrenewable resources.

12. Classify each use as either recycling or reusing:

    a. storing water in used juice bottles for an emergency
    b. converting plastic milk containers into fibers used to weave clothing fabric
    c. packing breakable items with shredded newspapers

Connecting the Concepts

13. Chromium minerals are found at three different mine sites in these forms:

    a. Site 1: Chromite, $FeCr_2O_4$
    b. Site 2: Crocoite, $PbCrO_4$
    c. Site 3: Chrome ochre, Cr_2O_3

    Based only on percent composition, at which site is chromium mining most feasible?

14. Earth has been compared to a space station.

    a. In what ways is this analogy useful?
    b. In what ways is it misleading?

15. Describe at least two benefits of discarding less waste material and recycling more of it.

16. One method of producing chromium metal includes, as the final step, the reaction of chromium(III) oxide with silicon at high temperature:

    $$2\,Cr_2O_3(s) + 3\,Si(s) \longrightarrow 4\,Cr(s) + 3\,SiO_2(s)$$

    a. How many moles of each reactant and product are shown in this chemical equation?
    b. What mass (in grams) of each reactant and product is specified by this equation?
    c. Show how this equation illustrates the law of conservation of matter.

Extending the Concepts

17. Why is aluminum metal more easily produced from recycled aluminum cans than from aluminum-containing materials such as clay, bauxite, or aluminum oxide ore?

18. In your laboratory experiences, you may encounter some compounds called hydrates. Examples of hydrates include $Na_2S_2O_3 \cdot 5H_2O$, $CaSO_4 \cdot 2H_2O$, and $Na_2CO_3 \cdot 10H_2O$.

    a. Why are these compounds called "hydrates"?
    b. Calculate the molar mass of each listed hydrate.
    c. Calculate the percent composition of water in each hydrate.
    d. Although hydrates contain significant amounts of water, they are found as solid substances at room temperature. How is this possible?

19. Atoms that presently make up your body may have once been part of *Tyrannosaurus rex*, Alexander the Great, or Cleopatra. Explain how this is possible.

MATERIALS: DESIGNING FOR PROPERTIES

SECTION D

As Earth's chemical resources continue to be extracted and used, issues of scarcity or cost, or other economic or political factors sometimes motivate a consideration of alternatives. One option is to find, modify, or create new materials to serve as substitutes. An ideal substitute satisfies three requirements: It is plentiful, it is inexpensive, and, of course, its useful properties match or exceed those of the original material. Alternative materials seldom meet these conditions completely. As a result, the benefits and burdens involved in each option must be weighed. This is true whether the design challenge involves a new coin, a better microwave oven, or a new computer storage device.

D.1 STRUCTURE AND PROPERTIES: ALLOTROPES

You have learned in your study of chemistry that elements are the fundamental building blocks of matter. You also know that each element—identified by a name, symbol, and particular populations of electrons and protons in its atoms—displays characteristic properties. For example, elements can be classified as metals, nonmetals, and metalloids. You are also aware that some elements, such as magnesium, are reactive and others, such as neon, are virtually inert. Given all of that, see if you can solve the following puzzle.

A CASE OF ELUSIVE IDENTITY   ChemQuandary 2

Imagine that you are given samples of three solids. You are told that each is a pure element.

You examine the first sample. It is a black solid that feels soft and a bit "greasy" to the touch. It leaves black marks when rubbed on the table. It is a useful lubricant and a fairly good conductor of electricity. A gram of the material sells for less than a penny.

You then inspect the second sample. It is a colorless, glasslike solid. It leaves deep scratch marks when rubbed on the table—in fact, it is among the hardest substances known. It is useful as an abrasive and as a coating for saw blades. It is a nonconductor of electricity. Depending on the quality of the solid piece, it can sell for $50 per gram or more than $20,000 per gram.

Finally you look at the third sample. It is a fine, powdery solid made up of the roundest molecules found in nature. Although samples of the substance are present in prehistoric layers of Earth's crust, it was only discovered in 1985. Although the price of this substance is dropping, in pure form—gram-for-gram—it is currently more expensive than gold.

These are, indeed, three distinctly different substances. Yet you are told that all three samples are exactly the same element. In other words, each is composed only of atoms of one particular element—no impurities, no mixtures, no compounds.

How can atoms of one element make up such different materials? What additional information do you need to explain this? What element is being described?

Before you attempt to answer these questions, it will helpful for you to know that chemists recognize the three samples just described as allotropes of the same element. **Allotropes** are two or more forms of an element that have distinctly different physical or chemical properties.

To be considered allotropes, the forms of the element must be in the same state—either solid, liquid, or gas. Solid iron (Fe) and molten iron have distinctly different properties, but because they are in different states, they are not allotropes.

It is possible that some information in ChemQuandary 2 provided enough clues for you to identify the element involved: If you guessed carbon (C), you are correct. The identities of each of the three carbon allotropes can now be revealed.

The first sample, composed of a soft, black solid, is the carbon allotrope known as graphite. You encounter graphite either directly or indirectly nearly every day. It is a major component of pencil "lead." Due in part to its conductive properties, it is also found in many batteries.

The second sample is carbon in the form of diamond. Its melting point is among the highest of any element. Both natural and synthetic diamonds exist. The cost of a diamond depends in part on its quality and optical characteristics. When carefully selected and cut, natural diamonds have high decorative value, even though their chemical formula, C(s), is the same as the formula for the graphite in common pencil lead.

The third sample is one of many recently identified allotropes of carbon, a group of ball-like and even tubelike structures known as fullerenes. The particular fullerene described here is buckminsterfullerene, a 60-carbon hollow sphere (C_{60}), which resembles a soccer ball. Other fullerene molecules with formulas such as C_{70}, C_{240}, and C_{540} are known, as are structures composed of "rolled up" layers of carbon atoms in the form of hollow tubes termed nanotubes.

How do chemists account for the different properties of allotropes? The explanation lies in how the atoms of the element are linked and organized—that is, in the structure of the substances. Although the three allotropes of carbon are all composed only of carbon atoms, they have distinctly different atomic arrangements. The three allotrope models in Figure 23 illustrate this fact.

Because graphite and diamond are both pure carbon, they can burn in oxygen to produce carbon dioxide gas and thermal energy. Despite that, diamonds are not recommended as a heating fuel!

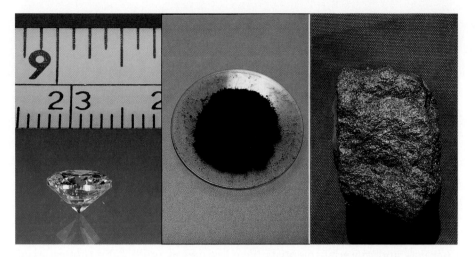

Figure 23 *Allotropic forms of carbon—their appearance and structures. From left to right, diamond, fullerene (in the form of buckminsterfullerene), and graphite. The structures shown of diamond and graphite are just small portions of three-dimensional structures that continue in every direction.*

 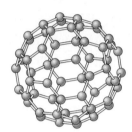 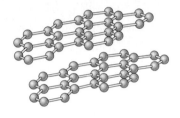

LINKING PROPERTIES TO STRUCTURE
Building Skills 11

 Structure and Properties: Allotropes

Based on the structural models shown in Figure 23, answer the following questions about some properties of carbon allotropes. The first question has been answered for you.

1. What feature of diamond's structure may account for its property as a hard, rigid substance, one that can scratch most other materials?

   Notice that every carbon atom in the interior of the diamond model is linked to four other carbon atoms by individual chemical bonds. These bonds have the effect of holding each carbon atom in place in a rigid three-dimensional structure. Each carbon-carbon linkage in diamond is a covalent bond involving the sharing of a pair of electrons. Thus "bending" or "denting" a diamond crystal would be extremely difficult because it would involve breaking a network of strong chemical bonds that locks each carbon atom in a fixed position. This accounts for diamond's rigidity and hardness.

2. How might the structure of diamond help explain why it is sometimes found in the form of large, single crystals?

3. What feature of graphite's structure might account for its usefulness as a lubricant?

4. Why are fullerenes "powdery" as solids rather than composed of large-scale "chunks"?

5. A molecule of buckminsterfullerene (C_{60}) can be regarded as a hollow sphere. Chemists have demonstrated that it is possible to place an atom of another element inside the sphere. Can you think of any practical application for "carrying" atoms of another element inside fullerene molecules?

D.2 MODIFYING PROPERTIES

Throughout history—first by chance and more recently guided by science—humans have greatly extended the array of materials available for their use. Chemists and material scientists have learned to modify the properties of matter by physically blending or chemically combining two or more substances. Sometimes only slight changes in a material's properties are desired. At other times chemists create new materials with properties dramatically different from those of the constituent substances.

Black "lead" in a pencil is mainly graphite, a natural form of the element carbon. You have just learned that pure carbon can display distinctively different properties depending on how the carbon atoms are bonded in three-dimensional space. Thus carbon as graphite displays properties that are uniquely different from those of carbon as diamond or carbon as fullerene.

If pencil lead is mainly graphite, however, why are the properties of hard pencil lead (such as No. 4), which produces very light lines on paper, so different from the properties of soft writing lead (such as No. 1), which makes very broad, easily smudged lines? As you might suspect, the properties of pencil lead are controlled by the amount of another material that is mixed with the graphite. That second material is clay. Increasing the quantity of clay mixed with graphite produces harder pencil lead because less graphite can be rubbed off onto paper.

Examples of other materials designed or modified to meet specific needs abound. Clay, one of the most plentiful materials on this planet, is composed mainly of silicon and oxygen atoms and aluminum ions, along with magnesium, sodium, and potassium ions and water molecules. Early humans found that clay mixed with water, then molded and heated, formed useful ceramic products such as pottery and bricks.

In more recent times, researchers have used newly developed techniques and materials to produce engineering **ceramics**. Figure 24 compares the sources, processing, and products of conventional ceramics with newer, stronger engineering ceramics.

What properties of conventional ceramics made them useful in pottery and bricks? Characteristics such as hardness, rigidity, low chemical reactivity, and resistance to wear were certainly important. The main attractions of ceramics for future use, however, are their high melting points and their strength at high temperatures. Indeed, such ceramics have become attractive substitutes for steel in some applications. For example, diesel or turbine engines made of ceramics can operate at higher temperatures than metal engines. Such high-temperature engines run with increased efficiency, thus reducing fuel use.

> When first discovered, graphite was mistakenly identified as a form of lead, and the name stuck.

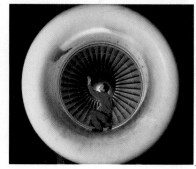

The major problem still facing researchers is that ceramics are brittle. They can fracture if exposed to rapid temperature changes, such as during hot engine cool-down. To avoid cracks in ceramic engine components (and resulting engine failure), scientists and engineers have developed precise manufacturing methods that carefully control the microstructure of the material. As these manufacturing practices become more economical, ceramics are likely to be more widely used in high-temperature applications.

Plastics have already replaced metals for many uses. These synthetic substances are composed of complex carbon-atom chains and rings with hydrogen and other atoms attached. Plastics generally are less dense and can be designed to be "springy" or resilient in situations where metals might become dented. Plastic automobile bodies are one example. The properties of certain plastics can be custom-made—in some cases without even changing the material's chemical composition. For example, polyethylene can be tailored to display relatively soft and pliable properties (as in a squeeze bottle for water) or crafted to be hard and brittle, almost glasslike in its behavior. Unfortunately, most plastics are made from petroleum, an important nonrenewable resource already in great demand as a fuel. You will learn more about petroleum in Unit 3.

Figure 24: *Conventional ceramic products such as bricks and pottery are made from clay. For example, kaolinite clay (top left) can be formed, by hand or machine, and fired in a kiln. Engineering ceramics, which may be higher melting, stronger, or less brittle than conventional ceramics, are produced from various minerals. They are designed for use in everything from the tiny resistors in computer circuit boards to huge jet turbines, both shown here.*

Optical fibers developed by chemists, physicists, and engineers are well on their way to replacing conventional copper wires in phone- and data-transmission lines. Voice or electronic messages are sent through these thin, specially designed glass tubes of very pure silicon dioxide (SiO_2) as pulses of laser light. As many as 50 000 phone conversations or data transmissions can take place simultaneously in one glass fiber the thickness of a human hair. A typical 72-strand optical fiber ribbon can carry well over a million messages. The fiber's larger carrying capacity and noise-free characteristics outweigh its higher initial cost.

Chemists, chemical engineers, and materials scientists continue to find new and better alternatives to traditional materials for a variety of applications. Such custom-tailoring at the molecular level is possible because of chemical knowledge—knowledge of how the atomic composition of materials affects their observable properties and behavior.

Now it's your turn to consider possible uses of some alternative materials.

ALTERNATIVES TO METALS   Building Skills 12

1. Select four uses of copper from the list found in Figure 11 (page 115). For each use, suggest an alternative material that could serve the same purpose. Consider both conventional materials and possible new materials.

2. Suggest some common metallic items that might be replaced by ceramic or plastic versions.

3. Suppose silver became as common and inexpensive as copper. In what uses would silver most likely replace copper? Explain.

D.3 STRIKING IT RICH   Laboratory Activity

Introduction

Seeing is believing—or so it is said. In this activity, you will change the appearance of some pennies through chemical and heat treatment.

Procedure

1. In your laboratory notebook, prepare a data table similar to the one shown here, leaving plenty of room for your observations.

DATA TABLE

Condition	Appearance
Untreated penny	
Penny treated with Zn and $ZnCl_2$	
Penny treated with Zn and $ZnCl_2$ and heated in burner flame	

2. Obtain three pennies. Use steel wool to clean each penny until it is shiny. Record the appearance of the pennies.

3. Set aside one of the clean pennies to serve as a **control**—an untreated sample that can be compared later to the other two treated coins.

4. Weigh a 2.0-g to 2.2-g sample of granulated zinc (Zn) or zinc foil. Place it in a 250-mL beaker.

5. Use a graduated cylinder to measure 25 mL of 1 M zinc chloride ($ZnCl_2$) solution. Add the solution to the beaker containing the zinc metal. **CAUTION:** *Zinc chloride solution can damage skin. If any accidentally spills on you, ask a classmate to notify your teacher immediately; wash the affected area with tap water immediately.*

6. Cover the beaker with a watch glass and place it on a hot plate. Gently heat the solution until it just begins to bubble, then lower the heat to continue gentle bubbling. Do not allow the solution to boil vigorously or become heated to dryness. **CAUTION:** *Note the warning about zinc chloride solution in Step 5.*

7. Using forceps or tongs, carefully lower two clean pennies into the solution in the beaker. To avoid causing a splash, do not drop the coins into the solution. Put the watch glass on the beaker and keep the solution boiling gently for two to three minutes. You will notice a change in the appearance of the pennies during this time.

8. Using forceps or tongs, remove the two coins from the beaker. Rinse them under running tap water, then gently dry them with a paper towel. Set one treated coin aside for later comparisons.

9. Briefly heat the other treated, dried coin in the outer cone of a burner flame, holding it with forceps or tongs, as shown in Figure 25. Heat the coin only until you observe a color change, which will take 10 to 20 seconds. Do not overheat.

10. Immediately rinse the heated coin under running tap water, and gently dry it with a paper towel. Record your observations.

11. Observe and compare the appearance of the three pennies. Record your observations.

12. When finished, discard the used zinc chloride solution and the used zinc as directed by your teacher.

13. Wash your hands thoroughly before leaving the laboratory.

Figure 25 *Heating the treated penny.*

Questions

1. a. Compare the color of the three coins—untreated (the control), heated in zinc chloride solution only, and heated in zinc chloride solution and in a burner flame.

   b. Do the treated coins appear to be composed of metals other than copper? If so, explain.

2. If someone claimed that a precious metal was produced in this activity, how would you decide whether the claim was correct?

3. Identify at least two practical uses for metallic changes similar to those you observed in this activity.

4. a. What happened to the copper atoms originally present in the treated pennies?

   b. Do you think the treated pennies could be converted back to ordinary coins? If so, what procedures would you use to accomplish this? **CAUTION:** *No laboratory work should be performed without your teacher's approval and direct supervision.*

(a)

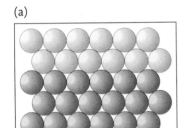

(b)

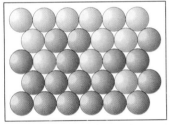

(c)

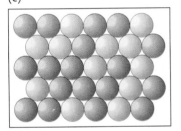

Figure 26 *Forming a brass outer layer on a penny. Brass samples contain mostly copper and between about 10% and 40% zinc (by mass).*

D.4 COMBINATIONS OF ELEMENTS: ALLOYS AND SEMICONDUCTORS

The laboratory activity you just completed demonstrated how metallic properties can be modified by creating an **alloy,** a solid combination of atoms of two or more metals. The immersion of a penny in hot zinc chloride solution produced a silvery alloy of zinc and copper called γ-brass. When you heated the penny in the burner flame, the zinc and copper atoms mixed more. The overall mixing of zinc and copper atoms to form brass is depicted in Figure 26. The resulting solid solution has a different concentration of zinc and copper and is known as α-brass. Brass has a golden color, unlike either copper or zinc. It is also harder than copper metal. Some other common alloys with familiar names are listed in Figure 27.

It is clear that one way to modify the properties of a particular metal is to form it into an alloy, just as you did when you produced a "gold-colored" penny. Often the results of alloying metals are unexpected, as you are about to discover.

FIVE CENTS' WORTH   ChemQuandary 3

A U.S. nickel coin is composed of an alloy of nickel and copper.

1. Based on your familiarity with that common five-cent coin, do you think the alloy in the coin contains more atoms of copper or more atoms of nickel? Why?

Now consider the facts: The nickel-coin alloy is composed of 25% nickel and 75% copper.

2. What does this suggest about a difference between an alloy (in this case a "solid solution" of copper and nickel atoms) and a simple mixture of powdered copper and powdered nickel?

In addition to alloys composed of "solid solutions" of two or more metals, other useful alloys include some that are well-defined compounds—that is, they have a constant, definite ratio of component metallic atoms. One example is Ni_3Al, a low-density, strong metallic alloy of nickel and

Common Alloy Compositions and Uses

Alloy and Composition	Examples (composition given in mass percent)	Comments
Brass Copper and zinc	Red brass 90% Cu, 10% Zn Yellow brass 67% Cu, 33% Zn Naval brass 60% Cu, 39% Zn, 1% Sn	Properties of brass vary with the proportion of copper and zinc and with the addition of small amounts of other elements. Brass is used for plumbing and lighting fixtures, rivets, screws, and ships.
Bronze Primarily copper with phosphorus, tin, zinc, and other elements	Coinage bronze 95% Cu, 4% Sn, 1% Zn Aluminum bronze 90% Cu, 10% Al Hardware bronze 89% Cu, 9% Zn, 2% Pb	Bronze is harder than brass. Its properties depend on the proportions of its components. Bronze is used for bearings, machine parts, telegraph wires, gunmetal, coins, medals, and bells.
Steel Primarily iron with carbon and small amounts of other elements	Steel 99% Fe, 1% C Nickel steel 96.5% Fe, 3.5% Ni Stainless steel 90–92% Fe, 0.4% Mn, <0.12% C, Cr (trace)	The properties of steel are often determined by the carbon content. High-carbon steel is hard and brittle; low- or medium-carbon steel can be welded and tooled. Steel is used for automobile and airplane parts, kitchen utensils, plumbing fixtures, and architectural decoration.
Other common alloys	Pewter 85% Sn, 6.8% Cu, 6% Bi, 1.7% Sb	Pewter is often used for figurines and other decorative objects.
	Mercury amalgams 50% Hg, 20% Ag, 16% Sn, 12% Cu, 2% Zn	Mercury amalgams have often been used for dental fillings.
	14 carat gold 58% Au, 14–28% Cu, 4–28% Ag	14-carat gold is popular in jewelry.
	White gold 90% Au, 10% Pd	White gold is also principally used for jewelry.

aluminum that is used as a component of jet aircraft engines. A very hard chromium-platinum alloy, Cr_3Pt, forms the basis of some commercial razor blade edges. And a special group of alloys, including the niobium-tin compound Nb_3Sn, displays superconductivity—the ability to conduct an electric current without any electrical resistance—if cooled to a sufficiently low temperature.

Figure 27 *Some common alloys and their uses.*

Combination of Elements: Alloys

Silicon Valley in California got its name from the element that is vital to the large number of computer-related industries located there.

Thus one important strategy for modifying the properties of metals is to produce alloys, which have properties that differ from those of the component elements. In fact, dramatic changes in the properties of a substance are possible when an extremely small amount of an element—as small as one atom per million atoms—is intermingled within a solid substance.

The metalloid known as silicon belongs in a class of materials called **semiconductors.** What does this term mean? You should recall that some elements (metals) are good conductors of electricity, whereas other elements (nonmetals) are not. Semiconductors, as the name implies, lie somewhere in between. In addition to silicon (Si), other elements and compounds that have semiconducting properties include germanium (Ge), tin (Sn), gallium arsenide (GaAs), and cadmium sulfide (CdS). Locate the elements identified as semiconductors on the Periodic Table (page 104). In what region of the table do they appear?

In the crystal structure of pure silicon, each atom is bonded to four other atoms. As a result, silicon's electrons are incapable of moving through the crystal. In other words, crystals of pure silicon display poor electrical conductivity at normal temperatures.

Although it sounds strange, adding certain impurities to pure silicon dramatically enhances its semiconductor properties. The impurities are atoms of other elements such as phosphorus, arsenic, aluminum, or gallium. This process, called **doping,** creates a situation in the solid silicon that allows charge carriers—either electrons or tiny regions of positive charge—to become mobilized within the crystal. This greatly improves silicon's semiconductor characteristics.

Most semiconductor devices (such as transistors and integrated circuits used in computers and other electronics) are based on solid crystals of silicon that contain intentionally added impurities. Those added atoms are responsible for the present silicon-based solid-state era, and, of course, for the existence of Silicon Valley itself.

Alloys and semiconductors involve changing the internal composition of a material to favorably affect its properties and uses. However, as you will now discover, there are other ways to modify or improve the properties of materials extracted from Earth's mineral resources.

D.5 MODIFYING SURFACES

In Section A you learned that skyrocketing copper prices in the early 1980s forced the United States to use zinc as the bulk metal in the penny. Copper was used to cover the coin's surface. This copper exterior not only preserves the appearance of the coin but also protects it from corrosion. The use of one material to protect the surface of another, less durable material is by no means new or unusual. A layer of protective material changes the properties of a manufactured product while allowing a less expensive or more available material to be used in the bulk of the item. As you continue your reading, you will learn about several types of surface treatments, both new and old, and how they can be used to enhance materials and products.

Coatings

The surface treatment that is probably most familiar to you is a **coating.** Coatings include paints, varnishes, and shellacs commonly applied to homes, cars, and many other products you buy and use. These materials are generally applied late in the manufacturing or construction process and are physically or chemically attached to the surface to be protected. Although coatings sink into pores in the base material, no bonding between the coating and the base takes place.

Paint, a typical example of a coating, consists of three components: pigment, solvent, and resin. The pigment provides color, the solvent allows application of the paint in liquid form, and the resin provides the desirable protective properties. Although paints have existed for centuries, growing concerns about the solvents released in their application and the heavy metals commonly found in their pigments have spurred the development of new formulations and methods of application. Of particular interest is the technique of powder coating, which eliminates the need for a solvent. See Figure 28. In this method, all components of the coating are blended and ground into an extremely fine powder. The product to be coated is then cleaned and sometimes pretreated. The powder is mixed with compressed air, pumped into spray guns, and given an electrical charge. The product is electrically grounded, allowing it to be coated by the charged powder. After coating, the product is heated to "cure" the powder into a tough shell. Many products, including bicycle and auto parts, are powder-coated to produce long-lasting, attractive surfaces.

> A formulation, as used in industry, is much like a recipe for a product.

Plating

The process used to cover zinc with copper in making coinage involves using direct-current (DC) electricity, which causes redox reactions to occur between the metal and a metal-ion solution. This general process is known as **electroplating.** As you have already learned, cations of most metallic elements can be reduced. This fact is exploited in electroplating. For example,

Figure 28 *This bicycle's metal parts have been powder-coated to resist corrosion.*

All That's Gold Doesn't Glitter . . .
Or Does It?

Did you ever wonder where the gold metal that makes up a high-school ring comes from? Or who is responsible for locating and recovering the material? **John Langhans** may have helped. He's a metallurgist with Barrick Goldstrike, a mining company in Elko, Nevada, that finds and recovers gold.

John's job is not an easy one. The normal concentration of gold in Earth's crust is only about 15 parts per billion (ppb). During the 1800s and early 1900s, miners looked for areas where gold was concentrated in the form of nuggets. More refined mining techniques enabled miners to locate deposits of finely divided gold. Such deposits account for most of today's gold mines.

Gold is often associated with certain types of rocks; quartz is an example. It is relatively easy to recover gold from these rock deposits. In the past decade, however, many of these deposits have been mined out, and the remaining gold is much harder to process.

When associated with sulfide minerals, gold is known as refractory ore. More sophisticated equipment and techniques must be employed to recover this gold. By using knowledge of chemistry and highly efficient recovery methods, John and his coworkers can extract gold from this material.

One method John's company uses is "autoclaving." In this process, the sulfide minerals are cooked under fairly high temperatures and pressures to dissolve the sulfides and expose the gold. The auto-claves work in a manner similar to that of a pressure cooker. You may be familiar with this piece of cookware if it is used to prepare meals in your home. Once the gold is exposed, a dilute basic solution containing cyanide ions (CN^-) is added to the ore to dissolve the gold. A gold cyanide complex $[Au(CN)_2]^-$ forms when gold atoms come in contact with the solution:

$$4 Au + 8 CN^- + 2 H_2O + O_2 \longrightarrow 4 [Au(CN)_2]^- + 4 OH^-$$

Later, the gold is recovered through a process called electrowinning. Electrowinning uses an electrical current to transfer the gold onto steel wool. Finally, technicians heat the gold-plated steel wool to liquefy the gold, and pour the liquid gold into bar molds.

John supervises laboratory personnel whose job it is to research and develop new methods for recovering gold that are both safe and economical. He and his team also monitor the processing operations of mines to ensure that the environment is not degraded as a result of those operations.

> John supervises laboratory personnel whose job it is to research and develop new methods for recovering gold. . .

Some Questions To Ponder . . .

Class rings are not 100% pure gold. Because pure gold is very soft, it is mixed with other metals (copper, silver, nickel, or zinc) to increase hardness and durability. The greater the amount of added metal, the lower the karat value. Pure gold is 24 karat. Most gold-containing rings are probably 12 karat, or about 50% gold.

Because of the specialized processes that Barrick Goldstrike uses, the company is able to profitably recover gold from refractory ore, even at grades as low as 5 grams gold per metric ton ore.

1. If a 12-karat-gold class ring has a mass of 11.2 g, how many metric tons of ore were processed in order to produce the gold to make the ring?

2. What is the value of the gold in the class ring if the market value of gold is about $10.50/g? (You may want to use the current actual price of gold, which is available in the business section of many newspapers or on the World Wide Web.)

3. During the 1800s, gold miners looked for gold in the form of nuggets. What property of gold allows it to be found in this form?

metal bumpers on trucks are often made of steel. In wet or snowy climates, the exposed steel would quickly corrode, or rust. Manufacturers protect the steel by coating it with chromium and nickel. See Figure 29.

Coins, too, have been plated with nickel for protection from corrosion. Writing half-reactions, which represent the reduction and oxidation parts of a redox process separately, can be helpful in understanding the plating process. The reduction half-reaction for plating nickel on steel can be written as follows:

$$Ni^{2+}(aq) + 2\,e^- \longrightarrow Ni(s)$$

But where do the electrons in this equation come from? Electroplating requires a power source, usually a battery or power supply when applied in the laboratory. The power source supplies electrons to the cathode (where reduction occurs) but it does not create those electrons. The ultimate source of the electrons in any electrochemical process is the anode. In electroplating, the anode is usually made of the metal to be plated. As metal cations are reduced, removed from the solution, and deposited on the object attached to the cathode, metal atoms at the anode are oxidized and dissolve into the plating solution. Figure 30 illustrates this process.

Thus the other half-reaction for the plating system is an oxidation, the reverse of the reaction shown above. Based on what you know about metal reactivities, would you expect this system of reactions to proceed spontaneously? Although systems that can plate metals without applied electric current have been developed, they must include a chemical reducing agent—which is sometimes more expensive than electric current.

Unlike coatings, platings are usually bonded to the surface of the bulk material through a metallic bond. This is desirable because it keeps the metal finish firmly attached to the object. After several layers of atoms have been deposited, the plating has the properties of the plating metal and can thus impart the properties for which it was selected. But what about the first few layers of atoms? Metallurgists have known for years that these layers tend to have properties distinctive from both the plating and the base materials. This knowledge helped lead to the development of another type of surface modification—thin films.

Figure 29 *Chrome plating can protect exposed steel surfaces.*

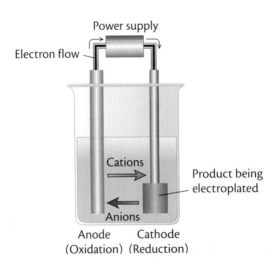

Figure 30 *The process of electroplating.*

Thin Films

Films are materials distinguished by the incredibly small thickness of their deposition on a surface. Modern thin films may be only one or two atoms or molecules thick! Films made of metal compounds, such as carbides and oxides, were originally developed to harden the edges of cutting tools. They are also used for decorative or protective purposes on many metal products. Alternating layers of these materials may enhance certain desired properties of the metal even more dramatically.

Thin metal films find many applications in the optics and electronics industries. For example, tungsten films interconnect circuits on microelectronic devices. Gold films can be deposited on glass or other surfaces to produce high-quality mirrors. Semiconductors also employ thin films in their construction.

Scientists are looking at the model provided by nature to develop new thin-films approaches. For example, seashells are created through the deposition of layers of inorganic and organic substances. Similar approaches are being used to create layers of engineering ceramics on surfaces.

The most common organic thin films are made of polymers, a subject you will explore in more detail in Unit 3. For now, however, it is important for you to know about the existence of polymer films because they have often been used to protect currency and important documents from counterfeiting and could be similarly used to protect coins. Many products you use daily are coated with polymers. Some common examples are automobile glass and the UV-protective and antiglare films on sunglasses. Why do you think it would be useful to have a thin film on automobile glass?

An interesting application of polymer thin films is in color-changing paint. Scientists have observed that the colors of butterflies result from the interference patterns of light caused by tiny fibers in their wings. Seeking to duplicate this effect, researchers have used both spun polyester and liquid-crystal polymers to develop car paint that appears to change color depending upon the angle from which it is viewed. Figure 31 illustrates this feature.

The development of thin films is a good example of the positive interaction of science and technology. As new discoveries are made, new applications result. Demands for better methods and more desirable properties lead to more basic research and subsequent applications. Because the properties of films are different from those of bulk materials, their study provides answers to important questions. The scope of their applications, meanwhile, continues to grow.

Figure 31 *The paint on this car changes color depending on the angle from which it is viewed.*

D.6 COPPER PLATING

Introduction

As you just learned, plating a metal requires the application of direct current, the presence of metal ions, and a suitable anode, usually made of the metal to be plated. In this activity, you will plate copper onto a nail. You will need the following materials:

- Beaker, U-tube, or transparent plastic (Tygon) tubing
- Copper plating solution
- ⚠ **CAUTION:** *Copper plating solutions are hazardous and corrosive.*
- Iron or zinc nail
- Copper metal strips
- 9-V battery or power supply
- ⚠ **CAUTION:** *Always be careful when working with electricity, especially high-voltage power supplies.*
- Wire leads with alligator clips
- Voltmeter or CBL kit (optional)

Procedure

Set up a system to deposit copper onto a nail. Keep in mind which metal should act as the anode and which as the cathode. Think about how electrons will flow. Test your setup. Record your results. If nothing happens, have your teacher check your setup. When you are finished with the activity, be sure to follow your teacher's instructions for disposal of wastes. Wash your hands thoroughly before leaving the laboratory.

Questions

Questions & Answers

1. What was the anode in this electrochemical cell?
2. Write an equation for the reaction that occurred at the anode.
3. What was the cathode in this cell?
4. Write an equation for the reaction that occurred at the cathode.
5. Does it matter what metal is used for the anode? Explain.
6. Does it matter what metal is used for the cathode? Explain.
7. Do you think this method would be useful for large-scale copper plating? Why or why not?

SECTION SUMMARY

Reviewing the Concepts

♦ **Allotropes of an element have distinctly different physical or chemical properties, variations explained by differences in the arrangement of the element's atoms.**

1. What is an allotrope?
2. Name two substances that are allotropes of the same element and compare their properties.

3. A diamond, a chunk of coal, and your pencil lead contain the same substance. How do they differ in properties? What accounts for the differences in value of the items?

♦ **An alloy possesses properties that differ, sometimes significantly, from the properties of its constituent elements.**

4. What is an alloy?
5. Give examples of two alloys you use every day.

6. What is an advantage of using an alloy rather than a pure metal for a particular purpose?

♦ **A poorly conductive material can sometimes be transformed into a semiconductor by adding a small amount of a particular impurity.**

7. Where are the majority of elements that behave as semiconductors located on the Periodic Table?

8. Would an unintended addition of a substance to a semiconductor be considered doping? Explain.

♦ **The surface properties of a material may be modified through application of coatings, thin films, or electroplated metals.**

9. Give three examples of how properties of materials can be modified by applying a coating.
10. Describe two ways in which electroplating is used in industry.

11. How do coatings, electroplatings, and thin films differ?
12. Can a rusty car bumper be protected from further rusting by electroplating? Explain.

Connecting the Concepts

13. Classify each major method of coating a surface as either a chemical or physical change. Explain your answers.
14. Use the activity series in Figure 16 (page 121) to decide which metals would be easiest to use as electroplating material.

15. New materials can successfully substitute for metals in some products. Explain why each of the following replacements can be made.
   a. Ceramics replace steel in turbine engines.
   b. Plastics replace steel in automobile bumpers.
   c. Optical fibers replace copper in phone wires.

Extending the Concepts

16. Obtain information about various allotropic forms of the element sulfur.
   a. Draw models depicting the major allotropes.
   b. Compare the properties of these allotropes.

17. Define *synergy*. How can this term be used to describe properties of alloys? Give two examples.
18. Use the Web and reference materials to find out why doping allows electrons to move more easily in semiconductors.

PUTTING IT ALL TOGETHER
Making Money

Choosing the Best Coin

Putting It All Together

When a new product is created, the design process often includes proposals from several individuals or teams. A panel of experts or consumers then chooses the "best" of the proposals. Because each school can submit only one design for the new coin, you will follow a similar process in choosing the best proposal. Each team will present its coin design and findings to the class, which will serve as the selection committee. The following paragraphs summarize the essential elements of your design presentation.

Coin Model

Include a model or detailed drawing of the proposed coin. If you create a model, it does not have to be made of the specified material but should resemble it. Ensure that all models and designs are at least five times the actual coin size. In addition, specify the actual size, thickness, and mass of the coin.

Materials Design

Describe the material or materials chosen for the coin. Present your rationale for selecting those materials. Include an analysis of both necessary and desirable properties of the chosen material(s), as well as methods to discourage counterfeiting.

Material Source Data

Provide details on the source of the raw materials needed to produce the new coin. Include a discussion of how the materials will be mined and/or processed. Estimate the long-term availability of those resources.

Life Cycle Analysis

Present an analysis of the life cycle of the proposed coin. How long is the coin expected to last in general circulation? Will material(s) making up the coin be recycled or reused? Consider the disposal or reuse of the material(s) in the coin as part of its production cost. (See next item.)

Cost Analysis

Present and analyze production costs of the new coin. These involve factors such as location of resources, mining, production or processing (including energy costs), and distribution. Will the cost of materials be less than the value of the coin itself? Are overall production costs reasonable based upon the expected lifetime of the coin?

Looking Back and Looking Ahead

As you come to the end of this unit, it is appropriate to pause and reflect on what you have learned thus far. You have discovered some of the working language of chemistry (symbols, formulas, and equations), laboratory techniques, and major ideas in chemistry (such as periodicity, the law of conservation of matter, and atomic-molecular theory). This knowledge can help you better understand some important societal issues. Central among these issues is the use and management of Earth's chemical resources, including water, metals, petroleum, food, and air.

You have also explored other issues that enter into policy decisions concerning technological problems and challenges. Although chemistry is often a crucial ingredient in recognizing and resolving such issues, many problems are far too complex for a simple technological "fix." Issues of policy are not usually "either/or" situations, but often involve many dimensions and considerations. As a voting citizen, you will be concerned with a variety of issues that require some scientific understanding. Tough decisions may be needed. The remaining units of this textbook will continue to prepare you for this important responsibility.

Unit 3 deals with petroleum, a nonrenewable chemical resource so important that it deserves its own unit.

UNIT 3

WHAT are the chemical and physical properties of hydrocarbons?

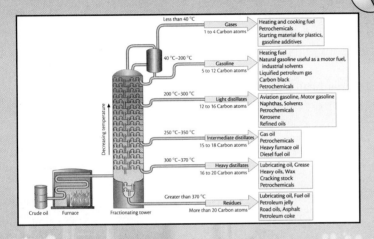

WHY do hydrocarbons make such good fuels?

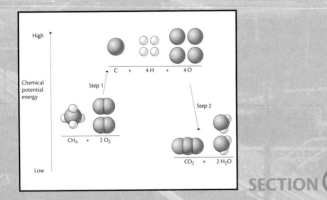

PETROLEUM: BREAKING AND MAKING BONDS

WHY are carbon-based molecules so versatile as chemical building blocks?

WHAT properties are important in considering substitutes for petroleum?

There is increased interest in alternative-energy transportation. Why? What are advantages and disadvantages to petroleum alternatives? How can the global supply of petroleum best be used? Turn the page to learn more about this energy-rich resource.

THE NEW ARL-600 TV COMMERCIAL—WORKING SCRIPT

Text/Script	Image/Sound
Clean . . .	(Video of driver and three passengers riding in the vehicle.)
Comfortable . . .	(The words "clean," "comfortable," "convenient" appear, in turn, as the announcer speaks them. "Clean" appears at the top of the screen; "Comfortable" appears in the middle; "Convenient" appears at the bottom. The "C" moves from the top to the bottom as the words appear.)
Convenient . . .	
Those three words describe the new breakthrough personal vehicle, the ARL-600.	(Show the ARL-600 driving along a road as announcer reads line.)
Clean . . .	("Clean" appears on the screen and moves from left to right.)
Unlike petroleum-fueled vehicles, the ARL-600 is emission-free—no air pollutants, smoke, or smog. Why? Because the ARL-600 is directly powered by electricity, thanks to its new, high-capacity electric storage batteries. You drive guilt-free, knowing you're helping, not hurting, the environment.	(Show the ARL-600 accelerating [after sitting at a stoplight] next to gasoline-burning vehicles to show the difference in emissions.)
Comfortable . . .	("Comfortable" appears on the screen and moves from the top left corner to the bottom right corner.)
You and three others can easily ride in the ARL-600. And—thanks to electrical power—the ride is so quiet that driving the ARL-600 seems more like a walk in the park.	(Show four people riding in the ARL-600, then change the perspective to that of the driver with the window rolled down. The driver can hear birds chirping and children playing as she drives past a park.)
Convenient . . .	("Convenient" appears on the screen and moves from top center to bottom center.)

Imagine an end to stops at busy gas stations. How? You simply recharge your ARL-600's batteries when you return home, and you're ready for another day of gasoline-free driving.	(The ARL-600 passes a gas station, pulls up to a house, and the driver "plugs in" the vehicle.)
When fully charged, your electric-powered ARL-600 will take you wherever you want to go within a 90-mile range—school, work, or even soccer practice.	(Images of the ARL-600 in a variety of settings—shopping center, office parking lot, school, recreation area.)
One place you *won't* need to visit, however, is the auto-repair shop. Fewer moving parts mean less time and money spent on engine repairs.	(The ARL-600 drives past an auto-repair shop.)
And using your petroleum-free ARL-600 is easy on the pocketbook—as little as two cents per mile to operate. Compare that to more than five cents per mile for gasoline-burning vehicles.	(Image of a piggy bank being shaken with coins [mostly pennies] falling out onto the table.)
Help conserve petroleum resources! Visit your ARL-600 dealer for a test drive today.	(Side view of an ARL-600 at a dealer showroom with several people examining it. The "ARL-600" logo then appears below the vehicle.)

You live in a world of new products, devices, and materials. Whether presented on a billboard, displayed on television, featured in a magazine, or announced on the radio, every advertisement attempts to sell its product by informing the audience about specific features. The product may be "faster," "lighter," "easier to use," "newer," or "great-tasting"—to sample just a few of many common product claims. The advertisement for the ARL-600 is no exception; it highlights several energy- and fuel-related features of a new and (so it is claimed) "petroleum-free" vehicle.

In this unit, you will gain the knowledge and perspective you need to analyze the merits of the claims made about the ARL-600. For example, you will learn what petroleum is, what chemical and physical properties make it so useful, and how it is used. Then you can evaluate whether the ARL-600 is actually "petroleum-free" or not.

But you will learn more about petroleum than its use as a fuel. Many products you use every day are made from petroleum—an outstanding example is plastics. What is it that allows petroleum to be useful as both a fuel to burn and a building block from which to make many new substances? How long will known world reserves of petroleum last? What alternatives to petroleum are there?

As you learn about this valuable resource, consider the energy and fuel claims made in the ARL-600 advertisement. You will soon be invited to analyze those claims. Later in the unit you will have the opportunity to produce a design for a new-vehicle advertisement based on the knowledge you have gained.

PETROLEUM— WHAT IS IT?

Introduction

The word "petroleum" is probably quite familiar to you. But do you know what petroleum is or what it is made of? Can you explain what properties make it useful for both burning and building? In this section you will explore the characteristics of some key compounds found in petroleum. Specifically, you will focus on their structure, bonding, and properties.

A.1 WHAT *IS* PETROLEUM?

> The word "petroleum" comes from the Latin words *petr-* ("rock") and *oleum* ("oil").

Petroleum is a vitally important world resource. As pumped from underground, petroleum is known as crude oil, or "black gold." This liquid varies from colorless to greenish-brown to black, and may be as thin as water or as thick as soft tar. Crude oil cannot be used in its natural state. Instead, it is shipped by pipeline, ocean tanker, train, or barge to oil refineries, where it is separated into simpler mixtures. Some of these mixtures are ready for use, whereas others require further refinement. Refined petroleum is chiefly a mixture of various **hydrocarbons**—molecular compounds that contain atoms of the elements hydrogen and carbon only. Can you see how this class of compounds got its name?

Nearly 50% of the total energy needs of the United States are met by burning petroleum. Thus most petroleum is consumed as a fuel. Converted to gasoline, petroleum powers millions of automobiles in the United States, each traveling an average of 11 000 miles annually. Other petroleum-based fuels provide heat to homes and businesses, deliver energy to generate electricity, and propel diesel engines and jet aircraft.

But petroleum's importance goes beyond its use as just a fuel. Its other major use is as a raw material from which a stunning array of familiar and useful products are manufactured—from CDs, sports equipment, clothing, automobile parts, and carpeting to prescription drugs and artificial limbs. Based on your experiences with petroleum fuels and products, what percent of petroleum would you estimate is used for burning? For building? Can you identify other uses of petroleum? The answers in the next paragraph may surprise you.

What did you predict for the percent of petroleum used for burning? Fifty percent? Sixty percent? Astonishingly, 84% of petroleum is burned outright as fuel. Only about 7% is used for producing substances such as medications and plastics. The remaining 9% is used as lubricants, road-paving materials, and an assortment of miscellaneous products. For every gallon of petroleum that is used to produce useful products, more than five gallons are burned to release energy.

What happens to molecules in petroleum when they are burned or used in manufacturing? As in all chemical reactions, the atoms become rearranged to form new molecules. When hydrocarbons burn, they react with oxygen gas in the air to form carbon dioxide (CO_2) gas and water vapor.

These gases disperse in the air. The hydrocarbon fuel is used up; it will take millions of years for natural processes to replace it. Thus petroleum is a nonrenewable resource—much like the minerals you studied in Unit 2.

Like other resources, petroleum is not uniformly distributed around the world. Approximately 57% of the world's known crude oil reserves are located in just five Middle Eastern nations: Iran, Iraq, Kuwait, Saudi Arabia, and the United Arab Emirates. By contrast, the petroleum reserves of North America amount to only about 7% of the world's known supply. The distribution of crude oil reserves does not necessarily correspond to population or use of petroleum. For example, Asia, the Far East, and Oceania account for 60% of the world's population, but this region has only about 4% of the world's petroleum reserves. Figure 1 shows these global distributions.

You have just learned what petroleum is, what it is used for, and where it is found. Petroleum is actually a complex mixture of hydrocarbons that must be refined or separated into simpler mixtures in order to be useful. In the following activity, you will find out about this basic separation process as you investigate a simple mixture of two liquids.

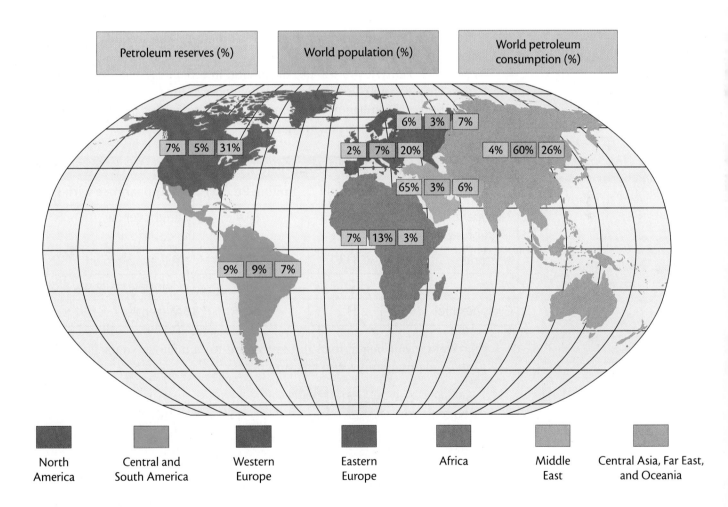

Figure 1 *Distribution of world's petroleum reserves, population, and consumption of petroleum.*

A.2 SEPARATION BY DISTILLATION

Introduction

You know that substances can often be separated by taking advantage of their different physical properties. One physical property commonly used to separate two liquids is density. But density will work only if the two substances are insoluble in each other, which is not the case with petroleum. Another physical property chemists often use is the boiling point of a substance. The separation of liquid substances according to their differing boiling points is called **distillation**.

As a liquid mixture is heated, the substance with the lower boiling point will vaporize first and leave the distillation flask; it will then be converted back to a liquid as it passes through a condenser—all before the second substance begins to boil. Each condensed liquid substance, called a **distillate**, can thus be collected separately.

As you might expect, heating the liquid mixture raises its temperature. However, once the first substance begins to boil and vaporize from the mixture, the temperature of the liquid remains steady until that component is completely distilled from the mixture. Continued heating will then cause the temperature to rise once again, this time until the second component begins to boil and distill.

The boiling points listed in Figure 2 are based on normal sea-level atmospheric pressure.

In this activity, you will use distillation to separate a mixture of two liquids. Then you will identify the two substances in the mixture by comparing the observed distillation temperatures with the boiling points of several possible compounds listed in Figure 2.

Figure 2 *Properties of substances that may be in the distillation mixture.*

Properties of Possible Components of Distillation Mixture			
Substance	Formula	Boiling Pt. (°C)	Appearance with I_2
2-Propanol (rubbing alcohol)	C_3H_7OH	82.4	Bright yellow
Acetone	C_3H_6O	56.5	Yellow to brown
Water	H_2O	100	Colorless to light yellow
Cyclohexane	C_6H_{12}	80.7	Magenta

Before you begin, read the Procedure to familiarize yourself with the intended observations and measurements.

Procedure

1. Construct suitable data tables to record your observations and measurements.

2. Assemble an apparatus similar to that shown in Figure 3. Label two beakers Distillate 1 and Distillate 2.

3. Using a clean, dry graduated cylinder, measure a 50-mL sample of the distillation mixture. Pour it into the flask and add a boiling chip.

Figure 3 *Distillation of a simple mixture.*

Water out

Thermometer   Glass tubing

Rubber
stopper

Cork
stopper

Condenser

125-mL
Florence
flask

Mixture to
be distilled

Hot plate

Water in

Distillate   Beaker

4. Record your observations of the starting mixture.

5. Connect the flask to a condenser as indicated in Figure 3. Ensure that the hoses are attached to the condenser and water supply as shown. Position the Distillate 1 beaker at the outlet of the condenser so it will catch the distillate.

6. Ensure that all connections are tight and will not leak.

7. Turn on the water to the condenser and then turn on the hot plate to start gently heating the flask.

8. Record the temperature at which the first drop of distillate enters the beaker. Then continue to record the temperature every 30 seconds. Continue to heat the flask and collect distillate until the temperature begins to rise again. At this point, replace the Distillate 1 beaker with the Distillate 2 beaker.

9. Continue heating and recording the temperature every 30 seconds until the second substance just begins to distill. Record the temperature at which the first drop of second distillate enters the beaker. Collect 1 to 2 mL of the second distillate. **CAUTION:** *Do not allow all of the liquid to boil from the flask.*

10. Turn off the heat and allow the apparatus to cool. While the apparatus is cooling, test the relative solubility of solid iodine (I_2) in Distillate 1 and Distillate 2 by adding a small amount of iodine to each beaker and stirring. Record your observations.

11. Disassemble and clean the distillation apparatus and dispose of your distillates as directed by your teacher.

12. Wash your hands thoroughly before leaving the laboratory.

Questions

1. Plot your data on a graph of time (*x* axis) vs. temperature (*y* axis). Be sure to include all features of a correctly drawn graph. See page 66.

2. a. Using your graph, identify the temperature at which the first substance distilled and the temperature at which the second substance distilled.

   b. Because the liquid temperature does not change appreciably during distillation of a particular component, those portions of the graph line should appear flat (horizontal). How well do these plateaus match the temperatures at which the first drops of each distillate were collected?

3. Using data in Figure 2 on page 178, identify each distillate sample.

4. Compile your data with the data of those students who distilled the same mixture.

   a. Find the mean and the mode for each distillate temperature.
   b. All laboratory teams did not obtain the same distillation temperatures. Why?

5. In which distillate was iodine more soluble? Explain.

6. What laboratory tests could you perform to decide whether the liquid left behind in the flask is a mixture or a pure substance?

7. Of the substances listed in Figure 2, which two would be most difficult to separate by distillation? Why?

8. a. What would a graph of time vs. temperature look like for the distillation of a mixture of all four substances listed in Figure 2?
   b. Sketch the graph and describe its features.

The statistical mode is the most frequently reported value in a set of data.

A.3 PETROLEUM REFINING

Unlike the simple laboratory mixture you investigated in the preceding activity, crude oil is a mixture of many compounds. Separating such a complex mixture requires the application of distillation techniques to large-scale oil refining. The refining process does not separate each compound contained in crude oil. Rather, it produces several distinctive mixtures called **fractions**. This process is known as **fractional distillation**. Compounds in each fraction have a particular range of boiling points and specific uses. Figure 4 illustrates the fractional distillation (fractionation) of crude oil.

First, the crude oil is heated to about 400 °C in a furnace. It is then pumped into the base of a distilling column (fractionating tower), which is usually more than 30 m (100 ft) tall. Many of the component substances of the heated crude oil vaporize. The temperature of the column is highest at the bottom and decreases toward the top. Trays are arranged at appropriate heights inside the column to collect the various fractions.

During distillation, the vaporized molecules move upward in the distilling column. The smaller, lighter molecules have low boiling points and either condense high in the column or are drawn off the top of the tower as gases. Fractions with higher boiling points contain larger molecules, which are more difficult to separate from one another and thus require more thermal energy to vaporize. These molecules condense in trays lower in the column. Substances with the highest boiling points never vaporize. These thick, or viscous, liquids—called bottoms—drain from the column's base. Each arrow in Figure 4 indicates the name of a particular fraction and its boiling-point range.

As you learn more about the characteristics of the fractions obtained from petroleum, think about how their products find uses in both traditional and electric vehicles.

Although the names given to various fractions and their boiling ranges may vary somewhat, crude oil refining always has the same general features.

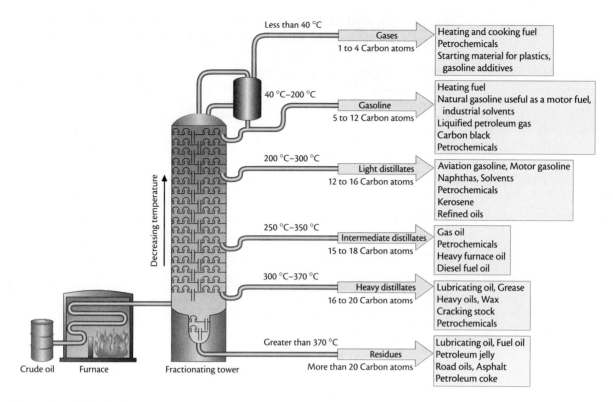

Figure 4   *A fractioning tower.*

A.4 A LOOK AT PETROLEUM'S MOLECULES

Petroleum's gaseous fraction contains compounds with low boiling points (less than 40 °C). These small hydrocarbon molecules, which contain from one to four carbon atoms, have low boiling points because they are only slightly attracted to each other or to other molecules in petroleum. Forces of attraction between molecules are called intermolecular forces. As a result of weak intermolecular forces, these small hydrocarbon molecules readily separate from each other and rise through the distillation column as gases.

Petroleum's liquid fractions—including gasoline, kerosene, and heavier oils—consist of molecules having from five to about twenty carbon atoms. Molecules with even more carbon atoms are found in the greasy solid fraction that does not vaporize. These thick, "sticky" compounds have the strongest intermolecular forces among all substances found in petroleum. It is not surprising that they are solids at room temperature.

Now complete the following activity to learn more about physical properties of hydrocarbons.

> Just as "interstate highway" means a road that runs between states, intermolecular forces means forces between molecules.

HYDROCARBON BOILING POINTS

Building Skills 1

Chemists often gather and analyze data about the physical and chemical properties of substances. These data can be organized in many ways, but the most useful techniques uncover trends or patterns among the data.

The development of the Periodic Table is an example of this approach. To refresh your memory about the Periodic Table, refer back to Section 2A.

Figure 5 *The boiling points of selected hydrocarbons.*

Hydrocarbon Boiling Points	
Hydrocarbon	**Boiling Point (°C)**
Butane	−0.5
Decane	174.0
Ethane	−88.6
Heptane	98.4
Hexane	68.7
Methane	−161.7
Nonane	150.8
Octane	125.7
Pentane	36.1
Propane	−42.1

In a manner similar to the one you used earlier to predict a property of an unknown element, you can examine patterns among the boiling points of some hydrocarbons in order to make valuable predictions. Use the data found in Figure 5 to answer the following questions.

1. a. In what pattern or order are Figure 5 data organized?
   b. Is this a useful way to present the information? Explain.

2. You are searching for a trend or pattern among these boiling points.
   a. Propose a more useful way to arrange these data.
   b. Reorganize the data table based on your idea.

Use your reorganized data table to answer these questions:

3. Which substance(s) are gases (have already boiled) at room temperature (22 °C)?

4. Which substance(s) boil between 22 °C (room temperature) and 37 °C (body temperature)?

5. What can you infer about intermolecular forces among decane molecules compared to those in butane?

A.5 CHEMICAL BONDING

Chemical Bonding

Hydrocarbons and their derivatives are the focus of the branch of chemistry known as **organic chemistry**. These substances are called organic compounds because early chemists thought that living systems—plants or animals—were needed to produce them. However, chemists have known for more than 150 years how to make many organic compounds without any assistance from living systems. In fact, starting materials other than petroleum can be used to produce organic compounds. You will learn about some of these starting materials in Section C.

In hydrocarbon molecules, carbon atoms are joined to form a backbone called a **carbon chain**. Hydrogen atoms are attached to the carbon backbone. Carbon's versatility in forming bonds helps to explain the abundance of different hydrocarbon compounds, as you will soon learn. Hydrocarbons

The carbon chain forms a framework to which a wide variety of other atoms can be attached.

can be regarded as "parents" of an even larger number of compounds that contain atoms of other elements attached to a carbon chain.

Electron Shells

How are atoms of carbon or other elements held to each other in compounds? The answer is closely related to the arrangement of electrons in atoms. You already know that atoms are made up of neutrons, protons, and electrons. Neutrons and protons are located in the small, dense, central region of the atom called the nucleus. Electrons occupy different **energy levels** in the space surrounding the nucleus. Similar energy levels are grouped into shells, each of which can hold only a certain maximum number of electrons. For example, the first shell surrounding the nucleus of an atom has a capacity of two electrons. The second shell can hold a maximum of eight electrons.

Consider an atom of helium (He), the first member of the noble-gas family. A helium atom has two protons (and two neutrons) in its nucleus and two electrons occupying the first, or innermost, shell. Because two is the maximum this shell can hold, the shell is completely filled.

The next noble gas, neon (Ne), has an atomic number of 10. This means that each neutral neon atom contains ten protons and ten electrons. Two electrons occupy (and fill) the first shell. The remaining eight electrons fill the second shell. In neon, each shell has reached its electron capacity.

Both helium and neon are chemically unreactive—their atoms do not combine with each other or with atoms of other elements to form compounds. By contrast, sodium (Na) atoms—with an atomic number of 11 and one more electron than neon atoms—are extremely reactive. Chemists explain sodium's reactivity as due to its tendency to lose that additional electron. Fluorine (F) atoms each have nine electrons—one less than neon atoms—and are also extremely reactive. Their reactivity is due to their tendency to gain an additional electron.

Noble-gas elements are essentially unreactive because their separate atoms already have filled electron shells. All but helium have eight electrons in their outer shells; helium needs only two to reach its first-shell maximum. A useful key to understanding the chemical behavior of many elements is to recognize that atoms with filled electron shells are particularly stable—that is, they are chemically unreactive. How does this guideline help to explain both the stability of noble-gas elements and the reactivity of elements such as sodium and fluorine?

When sodium metal reacts, sodium ions (Na^+) form. The $+1$ electrical charge indicates that each sodium atom has lost one electron. Each Na^+ ion contains eleven positively charged protons but only ten negatively charged electrons—thus the net $+1$ charge. With ten electrons, Na^+ possesses filled electron shells (two electrons in the first shell and eight in the second), just like a neon atom. Unlike sodium atoms, Na^+ ions are highly stable. In fact, the world's entire natural supply of sodium is found as Na^+ ions.

Fluorine atoms react to form fluoride ions (F^-). The -1 electrical charge indicates that each electrically neutral fluorine atom has gained one electron. Each fluoride ion contains nine protons and ten electrons—a net -1 charge. The ten electrons in an F^- ion constitute the same electron population found in a neon atom. Once again, an element has reacted to attain the special stability associated with filled electron shells.

> By losing an electron, each sodium atom has been oxidized. If you need to review this concept from Unit 2, see page 193.

> By gaining an electron, each fluorine atom has been reduced. See Unit 2, page 193, for a review of this concept.

Covalent Bonds

As you have just learned, electrons are either lost or gained in the formation of ionic substances. In molecular (non-ionic) substances, atoms achieve filled electron shells by sharing electrons rather than by losing or gaining electrons. Many molecular substances are composed of atoms of nonmetals that do not readily lose electrons. As you will see, the sharing of electrons between two nonmetallic atoms allows both atoms to complete their outer shells.

A hydrogen molecule (H_2) provides a simple example of electron sharing. Each hydrogen atom contains only one electron, so one more electron is needed to fill the first shell. Two hydrogen atoms can accomplish this if they each share their single electron. If an electron is represented by a dot ($\cdot$), then the formation of a hydrogen molecule can be depicted this way:

$$H\cdot + \cdot H \longrightarrow H\!:\!H$$

> The number of outer-shell electrons for any Group A element is equal to its group number in the Periodic Table. Thus Group 6 A elements each have six outer-shell electrons. The number of electrons in this section is also equal to the last digit in the 1–18 system.

The chemical bond formed between two atoms that share a pair of electrons is called a **single covalent bond**. Through such sharing, both atoms achieve the stability associated with complete electron shells. A carbon atom, atomic number 6, has six electrons—two in the first shell and four in the second shell. Only the electrons in the outer shell participate in chemical reactions. To fill the second shell to its capacity of eight, four more electrons are needed. These electrons can be obtained through covalent bonding.

Consider the simplest hydrocarbon molecule, methane (CH_4). In this molecule, each hydrogen atom shares its single electron with the carbon atom. Similarly, the carbon atom shares one of its four outer-shell electrons with each hydrogen atom. This arrangement is represented below.

$$4\,H\cdot \ + \ \cdot \overset{\textstyle\cdot}{\underset{\textstyle\cdot}{C}}\cdot \ \longrightarrow \ H\!:\!\overset{\textstyle H}{\underset{\textstyle H}{\overset{\cdot\cdot}{\underset{\cdot\cdot}{C}}}}\!:\!H$$

> Electron-dot formulas are also referred to as Lewis structures. G.N. Lewis is given credit for laying the foundation of our current understanding of bond formation.

Here, as in the formula for a hydrogen molecule, dots surrounding each element's symbol represent the outer-shell electrons for that atom. Structures such as these are called **electron-dot formulas**. The two electrons in each covalent bond "belong" to both bonded atoms. Dots placed between the symbols of two atoms represent electrons that are shared by those atoms.

When determining the number of electrons associated with each atom, each shared electron in a covalent bond is "counted" twice, once for each element. For example, count the dots surrounding each atom in methane. You should notice that each hydrogen atom has a filled outer electron shell—two electrons in its first shell. The carbon atom also has a filled outer electron shell—eight electrons. Each hydrogen atom is associated with one pair of electrons; the carbon atom has four pairs of electrons, or eight electrons.

For convenience, each pair of electrons in a covalent bond can be represented by a line drawn between the symbols of each atom. This yields another common representation of a covalently bonded substance called a **structural formula**.

$$H:\overset{\displaystyle H}{\underset{\displaystyle H}{C}}:H$$

Electron-dot formula
of methane, CH_4

$$H-\overset{\displaystyle H}{\underset{\displaystyle H}{C}}-H$$

Structural formula
of methane, CH_4

Although you can draw two-dimensional pictures of molecules on flat paper, assembling three-dimensional models gives a more accurate representation. Such an atomic model helps to predict a molecule's physical and chemical properties. The following activity provides an opportunity for you to assemble such models.

A.6 MODELING ALKANES   Laboratory Activity

Introduction

In this activity you will assemble models of several simple hydrocarbons. Your goal is to associate the three-dimensional shapes of these molecules with the names, formulas, and pictures used to represent them on paper.

Two types of molecular models are shown in Figure 6. Most likely, you will use ball-and-stick models. Each ball represents an atom, and each stick represents a single covalent bond (a shared electron pair) connecting two atoms. But of course molecules are not composed of ball-like atoms located at the ends of sticklike bonds. Experimental evidence shows that atoms are in contact with each other, much like what you see in space-filling models. However, ball-and-stick models are still useful because they can clearly represent the structure and geometry of molecules.

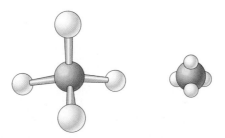

Figure 6   *Three-dimensional CH_4 models: ball-and-stick (left); space-filling (right).*

Look again at the electron-dot structure and structural formula for methane (CH_4) above. Methane, the simplest hydrocarbon, is the first member of a series of hydrocarbons known as **alkanes**. You will explore alkanes in this activity. Each carbon atom in an alkane forms single covalent bonds with four other atoms. Because each carbon atom is bonded to the maximum number of other atoms (four), alkanes are considered **saturated hydrocarbons**.

Procedure

1. Assemble a model of methane (CH_4). Compare your model to the electron-dot and structural formulas on page 185. Note that the angles defined by bonds between atoms are not 90°, as you might think by looking at the structural formula. If you were to build a close-fitting box to surround a CH_4 molecule, the box would be shaped like a triangular pyramid, or a pyramid with a triangle as a base. A **tetrahedron** is the name given to this three-dimensional shape.

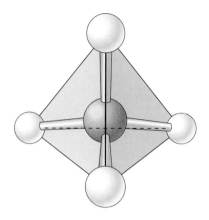

Figure 7 *The tetrahedral shape of a methane molecule.*

Why would the shape of a methane molecule be tetrahedral? Assume that the four pairs of electrons in the bonds surrounding the carbon atom—all with negative charges—repel one another. That is, the electron pairs stay as far away from one another as possible, arranging themselves so that they point to the corners of a tetrahedron. The angle formed by each C—H bond is 109.5°, a value that has been verified with several experimental methods. The angles are not 90°, as they would be if methane were flat. Verify this shape for yourself by arranging the atoms in your model.

2. Compare your three-dimensional model of methane to the representation of a tetrahedral molecule in Figure 7.

   a. How does the two-dimensional drawing (Figure 7) incorporate features that aid in visualizing the three-dimensional structure?
   b. Are there features of the two-dimensional figure that are difficult to translate into a three-dimensional structure? Explain.
   c. Translate your three-dimensional model into a two-dimensional drawing. Your drawing should convey the tetrahedral structure of methane.

3. Assemble models of a two-carbon and a three-carbon alkane molecule. Recall that each carbon atom in an alkane is bonded to four other atoms.

   a. How many hydrogen atoms are present in the two-carbon alkane?
   b. How many hydrogen atoms are present in the three-carbon alkane?
   c. Draw a ball-and-stick model, similar to the one in Figure 6 on page 185, of the three-carbon alkane.

4. a. Draw electron-dot and structural formulas for the two- and three-carbon alkanes.
   b. The molecular formulas of the first two alkanes are CH_4 and C_2H_6. What is the molecular formula of the third?

Examine your three-carbon alkane model and the structural formula you drew for it. Note that the middle carbon atom is attached to two hydrogen atoms, but the carbon atom at each end is attached to three hydrogen atoms. This molecule can be represented as CH_3—CH_2—CH_3, or $CH_3CH_2CH_3$. Formulas such as these provide convenient information about how atoms are arranged in molecules. For many purposes, such "condensed" formulas are more useful than molecular formulas such as C_3H_8.

Consider the formulas of the first few alkanes: CH_4, C_2H_6, and C_3H_8. Given the pattern represented by that series, try to predict the formula of the four-carbon alkane. If you answered C_4H_{10}, you are correct! The general molecular formula of all alkane molecules can be written as C_nH_{2n+2}, where

n is the number of carbon atoms in the molecule. So even without assembling a model, you can predict the formula of a five-carbon alkane: If $n = 5$, then $2n + 2 = 12$, and the formula is C_5H_{12}.

n can be any positive integer.

5. Using the general alkane formula, predict molecular formulas for the rest of the first ten alkanes. After doing this, compare your molecular formulas with the formulas given in Figure 8 to check your predictions.

The names of the first ten alkanes are also given in Figure 8. As you can see, each name is composed of a prefix, followed by -*ane* (designating an alk*ane*). The prefix indicates the number of carbon atoms in the backbone carbon chain. To a chemist, *meth*- means one carbon atom, *eth*- means two, *prop*- means three, and *but*- means four. For alkanes with five to ten carbon atoms, the prefix is derived from Greek—*pent*- for five, *hex*- for six, and so on.

6. Write structural formulas for butane and pentane.

7. a. Name the alkanes with these condensed formulas:
   (i) $CH_3CH_2CH_2CH_2CH_2CH_2CH_3$
   (ii) $CH_3CH_2CH_2CH_2CH_2CH_2CH_2CH_2CH_3$
   b. Write molecular formulas for the two alkanes in Question 7a.

8. a. Write the formula of an alkane containing 25 carbon atoms.
   b. Did you write the molecular formula or the condensed formula of this compound? Why?

9. Name the alkane having a molar mass of
   a. 30 g/mol.
   b. 58 g/mol.
   c. 114 g/mol.

Name	Number of Carbons	Alkane Molecular Formulas	
		Short Version	Long Version
Methane	1	CH_4	CH_4
Ethane	2	C_2H_6	CH_3CH_3
Propane	3	C_3H_8	$CH_3CH_2CH_3$
Butane	4	C_4H_{10}	$CH_3CH_2CH_2CH_3$
Pentane	5	C_5H_{12}	$CH_3CH_2CH_2CH_2CH_3$
Hexane	6	C_6H_{14}	$CH_3CH_2CH_2CH_2CH_2CH_3$
Heptane	7	C_7H_{16}	$CH_3CH_2CH_2CH_2CH_2CH_2CH_3$
Octane	8	C_8H_{18}	$CH_3CH_2CH_2CH_2CH_2CH_2CH_2CH_3$
Nonane	9	C_9H_{20}	$CH_3CH_2CH_2CH_2CH_2CH_2CH_2CH_2CH_3$
Decane	10	$C_{10}H_{22}$	$CH_3CH_2CH_2CH_2CH_2CH_2CH_2CH_2CH_2CH_3$

Some Members of the Alkane Series

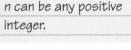

Modeling Alkanes

Figure 8 *The first ten alkanes.*

TRENDS IN ALKANE BOILING POINTS

In the laboratory activity involving distillation, you used a technique that separates liquid mixtures according to boiling points of substances. You also know that the fractions of petroleum are separated based on their boiling points. Why do the fractions with the highest boiling points contain the largest molecules? Why are the smallest molecules found in the fractions with the lowest boiling points? In this activity, you will explore this trend in alkane boiling points.

Using data for the alkanes found in Figure 5 (page 182) and Figure 8 (page 187), prepare a graph of boiling points. The x axis scale should range from 1 to 13 carbon atoms (even though you will initially plot data for 1 to 10 carbon atoms). The y axis scale should extend from $-200\,°C$ to $+250\,°C$.

1. Plot the data points. Draw a best-fit line through your data points according to these guidelines:

   - The line should follow the trend of your data points.
   - The data points should be equally distributed above and below the line.
   - The line should not extend past your data points.

   Figure 9 shows an example of a best-fit line.

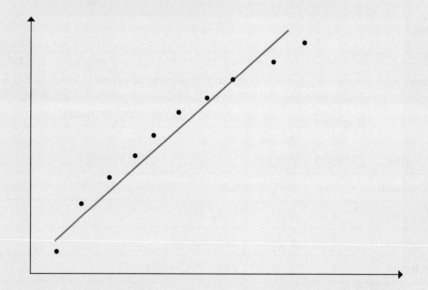

Figure 9 *An example of a best-fit line drawn through several data points.*

2. Estimate the average change in boiling point (in °C) when one carbon atom and two hydrogen atoms (—CH$_2$—) are added to a particular alkane chain.

3. The pattern of boiling points among the first ten alkanes allows you to predict boiling points for other alkanes.

a. Using your graph, estimate the boiling points of undecane ($C_{11}H_{24}$), dodecane ($C_{12}H_{26}$), and tridecane ($C_{13}H_{28}$). To do this, extend the trend of your graph line by drawing a dashed line from the graph line you drew for the first ten alkanes. This procedure is called **extrapolation**. Then read your predicted boiling points for C_{11}, C_{12}, and C_{13} alkanes on the *y* axis.

b. Compare your predicted boiling points to actual values provided by your teacher.

4. You learned that a substance's boiling point depends in part on its intermolecular forces, or attractions among its molecules. For the alkanes you have studied, what is the relationship between these attractions and the number of carbon atoms in each molecule?

CD-ROM WWW. **Intermolecular Forces and Boiling Point**

A.7 ALKANES REVISITED

Laboratory Activity

Introduction

The alkane molecules you have considered so far are **straight-chain alkanes**—each carbon atom is linked to only one or two other carbon atoms. In alkanes with four or more carbon atoms, many other arrangements of carbon atoms are possible. Alkanes in which one or more carbon atoms are linked to three or four other carbon atoms are called **branched-chain alkanes**. An alkane with four or more carbon atoms can have either a straight-chain or a branched-chain structure. In this activity you will use ball-and-stick molecular models to investigate such variations in alkane structures—variations that can lead to different properties.

A straight-chain structure:

C—C—C—C—C

A branched-chain structure:

C—C—C—C
      |
      C

Procedure

1. Assemble a ball-and-stick model of a molecule with the formula C_4H_{10}. Compare your model with those built by others. How many different arrangements of atoms in the C_4H_{10} molecule can be constructed?

Molecules that have identical molecular formulas but different arrangements of atoms are called **isomers**. By comparing models, convince yourself that there are only two isomers of C_4H_{10}. The formation of isomers helps to explain the very large number of compounds that contain carbon chains or rings.

2. a. Draw an electron-dot formula for each C_4H_{10} isomer.
   b. Write a structural formula for each C_4H_{10} isomer.

3. As you might expect, alkanes containing larger numbers of carbon atoms also have larger numbers of isomers. In fact, the number of different isomers increases rapidly as the number of carbon atoms

increases. For example, chemists have identified three pentane (C_5H_{12}) isomers. Their structural formulas are shown in Figure 10. Try building these and other models. Are other pentane isomers possible?

Alkane Isomers		
Alkane	Structural Formula	Boiling Point (°C)
C_5H_{12} isomers	$CH_3-CH_2-CH_2-CH_2-CH_3$	36.1
	$CH_3-CH-CH_2-CH_3$ $\quad\quad\;\; \mid$ $\quad\quad\; CH_3$	27.8
	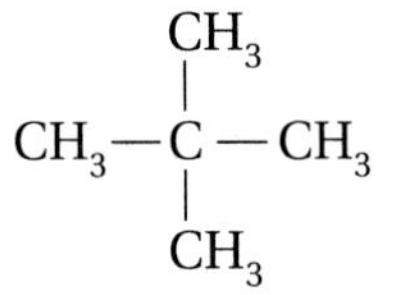 $\quad\quad\quad CH_3$ $\quad\quad\quad \mid$ CH_3-C-CH_3 $\quad\quad\quad \mid$ $\quad\quad\quad CH_3$	9.5
Some C_8H_{18} isomers	$CH_3-CH_2-CH_2-CH_2-CH_2-CH_2-CH_2-CH_3$	125.6
	$CH_3-CH_2-CH_2-CH_2-CH_2-CH-CH_3$ $\quad\quad\quad\quad\quad\quad\quad\quad\quad\quad\;\; \mid$ $\quad\quad\quad\quad\quad\quad\quad\quad\quad\quad CH_3$	117.7
	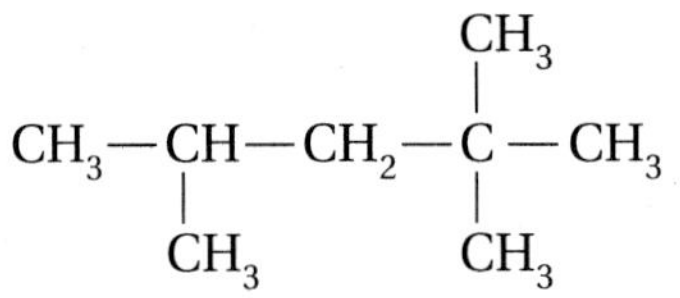 $\quad\quad\quad\quad\quad\quad\quad\quad CH_3$ $\quad\quad\quad\quad\quad\quad\quad\quad \mid$ $CH_3-CH-CH_2-C-CH_3$ $\quad\quad\; \mid \quad\quad\quad\quad \mid$ $\quad\quad CH_3 \quad\quad\; CH_3$	99.2

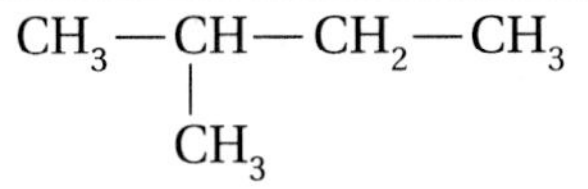

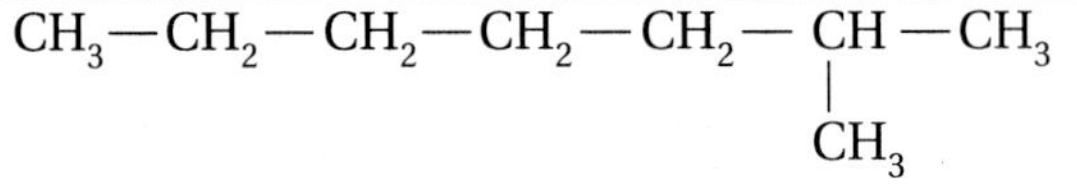

Figure 10 *Some pentane and octane isomers.*

4. Now consider possible isomers of C_6H_{14}.

   a. Working with a partner, draw structural formulas for as many different C_6H_{14} isomers as possible. Compare your structures with those drawn by other groups.

   b. How many different C_6H_{14} isomers were found by your class?

5. Build models of one or more C_6H_{14} isomers, as assigned by your teacher.

   a. Compare the three-dimensional models built by your class with corresponding structures drawn on paper.

   b. Based on your examination of the three-dimensional models, how many different C_6H_{14} isomers are possible?

Because each isomer is a different substance, it has its own characteristic properties. In the next activity, you will examine boiling-point data for some alkane isomers.

BOILING POINTS OF ALKANE ISOMERS

You have already observed that boiling points of straight-chain alkanes are related to the number of carbon atoms in their molecules. Increased inter-molecular forces are associated with the greater molecule-to-molecule contact possible for larger alkanes. Now consider the boiling points of some isomers.

1. Boiling points for two sets of isomers are listed in Figure 10 (page 190). Within a given set, how does the boiling point change as the extent of carbon-chain branching increases? Assign each of the following boiling points to the appropriate C_7H_{16} isomer: 98.4 °C, 92.0 °C, 79.2 °C.

    a. $CH_3-CH_2-CH_2-CH_2-CH_2-CH_2-CH_3$

    b.
    $$CH_3-CH_2-\underset{\underset{CH_3}{|}}{CH}-CH_2-CH_2-CH_3$$

    c.
    $$CH_3-CH_2-CH_2-\underset{\underset{CH_3}{|}}{\overset{\overset{CH_3}{|}}{C}}-CH_3$$

3. Here is the structural formula of a C_8H_{18} isomer:

    $$CH_3-CH_2-CH_2-\underset{\underset{CH_3}{|}}{\overset{\overset{CH_3}{|}}{C}}-CH_2-CH_3$$

    a. Compare it to each C_8H_{18} isomer listed in Figure 10. Predict whether it has a higher or lower boiling point than each of the other C_8H_{18} isomers.
    b. Would the C_8H_{18} isomer shown here have a higher or lower boiling point than each of the three C_5H_{12} isomers shown in Figure 10?

4. How do you explain the boiling point trends that you observed in this activity?

Chemists and chemical engineers use information about molecular structures and boiling points to separate the complex mixture known as petroleum into a variety of useful substances, many of which you are quite familiar with. In Section B, you will learn how bonding helps to explain the use of petroleum as a fuel.

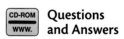 **Questions and Answers**

SECTION SUMMARY

Reviewing the Concepts

♦ **Petroleum (crude oil), a nonrenewable resource that must be refined prior to use, consists of a complex mixture of hydrocarbon molecules.**

1. What is a hydrocarbon?

2. What does it mean to refine a natural resource?

3. What is meant by saying that oil is "crude"?

4. What is the likelihood of discovering a pure form of petroleum that can be used directly as it is pumped from the ground? Explain your answer.

♦ **Petroleum is a source of fuels that provide thermal energy. It is also a source of raw materials for the manufacture of many familiar and useful products. On average, the United States uses about 18 million barrels of petroleum daily.**

5. About 16% of the petroleum used in the United States is used for "molecule building," producing nonfuel products that have a significant impact on everyday life. The remaining 84% of the petroleum is burned as fuel.

   a. What is the average number of barrels of petroleum used daily in the United States for building (nonfuel) purposes?
   b. How many barrels of petroleum are burned as fuel daily in the United States?

   c. List four household items made from petroleum.
   d. What materials could be substituted for each of these four items if petroleum were not available to make them?

6. Name several fuels obtained from crude petroleum.

7. List several products that might not be widely and easily available if petroleum supplies were to dwindle.

♦ **Liquid substances can often be separated according to their differing boiling points in a process called distillation.**

8. Rank the following hydrocarbons from their lowest boiling point to their highest: hexane (C_6H_{14}), methane (CH_4), pentane (C_5H_{12}), and octane (C_8H_{18}). Explain your rankings.

9. Sketch the basic setup for a laboratory distillation. Label the key parts.

10. Simple distillation is never sufficient to completely separate two liquids. Explain.

11. Explain why thermal energy is added at one point and removed at another point in the process of distillation.

♦ **Fractional distillation of crude oil produces several distinctive and usable mixtures called fractions. Each fraction contains molecules of similar sizes and boiling points.**

12. How does a fractional distillation differ from a simple distillation?

13. Petroleum fractions include light, intermediate, and heavy distillates and residues. List three useful products derived from each of these fractions.

14. Where in a distillation tower—top, middle, or bottom—would you expect the fraction with the highest boiling point range to be removed? Why?

15. After fractional distillation, each fraction is still a mixture. What must be done to further separate the components of each fraction?

♦ Hydrocarbon molecules, whose atoms are joined by covalent bonds, can be represented by electron-dot, structural, or molecular formulas.

16. What does each dot in an electron-dot diagram represent?

17. Each carbon atom has six electrons. Why does the electron-dot representation of carbon only show four dots?

18. Define the term "covalent bond."

19. Draw an electron-dot diagram for a branched six-carbon hydrocarbon.

20. a. What additional information does a structural formula convey that a molecular formula does not?

b. In what ways is a structural formula an inadequate representation of a real molecule?

♦ Alkanes, saturated hydrocarbons with single covalent bonds, have the general formula C_nH_{2n+2}.

21. Use the general molecular formula to write the molecular formula for an alkane containing

  a. 6 carbons.   b. 10 carbons.
  c. 16 carbons. d. 25 carbons.

22. Alkanes are said to be saturated hydrocarbons. What does "saturated" mean?

23. What does *-ane* imply about the bonding arrangement in compounds such as hexane, butane, methane, and octane?

♦ Isomers are molecules with identical molecular formulas but different arrangements of atoms. Each isomer is a separate substance with its own characteristic properties.

24. Draw structural formulas for at least three isomers of C_9H_{20}.

25. What is the shortest-chain alkane that can demonstrate isomerism?

26. a. Draw two hexane isomers—one straight chain and one branched chain.

b. Which of the two would have the lower boiling point? Explain your choice.

Connecting the Concepts

27. Why is petroleum considered a nonrenewable resource?

28. In a fractionating tower, petroleum is generally heated to 400 °C. What would happen if it were heated to only 300 °C?

29. The molar masses of methane (16 g/mol) and water (18 g/mol) are similar. At room temperature, methane is a gas and water is a liquid. Explain this difference.

30. The traditional unit of volume for petroleum is one barrel, which contains 42 gallons. Assume

that those 42 gallons provide 21 gallons of gasoline. How many barrels of petroleum does it take to operate an automobile for a year, assuming the auto travels 10 000 miles and goes 27 miles on a gallon of gas?

31. Which mixture would be easier to separate by distillation—a mixture of pentane and octane or a mixture of pentane and a branched-chain octane isomer? Explain the reasoning behind your choice.

Extending the Concepts

32. Is it likely that the composition of crude oil in Texas is the same as that of crude oil in Kuwait? Explain your answer.

33. Gasoline's composition, as blended by oil companies, varies in different parts of the nation.

a. Does the composition relate to the time of year?
b. If so, what factors help to determine the composition of gasoline in various seasons?

34. What kind of petroleum trade relationship would be expected between North America and the Middle East? If other world regions become more industrialized and global petroleum supplies decrease, how might the North America–Middle East trade relationship change?

35. The hydrocarbon boiling points listed in Figure 5 (page 182) were measured under normal atmospheric conditions. How would those boiling points change if atmospheric pressure were increased? (*Hint:* Although butane is stored as a liquid in a butane lighter, it escapes through the lighter nozzle as a gas.)

36. The two isomers of butane have different physical properties, as illustrated by their different boiling points. They also have different chemical properties. Explain how isomerism may contribute to differences in chemical behavior.

37. What properties of petroleum make it an effective lubricant?

38. When 1,2-ethanediol (ethylene glycol, also known as permanent antifreeze) is dissolved in water in an automobile's radiator, it helps keep the water from freezing. The permanent antifreeze-water solution has a lower freezing point than does pure water. Similarly, when an ionic substance such as table salt (NaCl) is dissolved in water, the solution freezes at a lower temperature than does pure water. Why is NaCl a highly undesirable additive for car radiators, whereas ethylene glycol is a suitable additive? (*Hint:* Compare the structure and chemical properties of these two substances.)

$$H-\underset{\underset{OH}{|}}{\overset{\overset{H}{|}}{C}}-\underset{\underset{OH}{|}}{\overset{\overset{H}{|}}{C}}-H$$

1,2-Ethanediol
(Ethylene glycol)

PETROLEUM AS AN ENERGY SOURCE

People have used petroleum for almost 5000 years. The first oil well was drilled in the United States in 1859 in Pennsylvania. Since then, human life has been greatly altered by the increasing use of petroleum. In the following activity, you will begin to appreciate just how much everyday life is influenced by petroleum's role as an energy source—not just in powering automobiles and other vehicles, but in energizing modern society itself.

USING FUEL ChemQuandary 1

Examine this textbook. It is a composite of a wide variety of materials—paper, colored inks, cardboard, coating material, and binding adhesive, to name a few. Indeed, a "materials story" could be written about this textbook—a story that would explore the origins of its component materials.

What is less obvious, perhaps, is that this textbook can also be analyzed in terms of energy: how energy (and the fuel from which the energy was produced) was involved in its manufacture, warehousing, and delivery to your school.

Complete an "energy trace" for your *ChemCom* textbook by following these steps. (a) Decide what general events must have occurred in the production of this textbook and its subsequent delivery to your school. (b) From where did the energy for each event come? In other words, what was the source of the needed energy? In your view, is the total cost of this textbook related more to the materials used (the paper, ink, adhesive, etc.) or to the energy required? Explain your reasoning.

B.1 ENERGY AND FOSSIL FUELS

Fossil fuels originate from biomolecules of prehistoric plants and animals. The energy released by burning these fuels represents energy originally captured from sunlight by prehistoric green plants during photosynthesis. Thus fossil fuels—petroleum, natural gas, and coal—can be thought of as forms of buried sunshine.

No one knows the exact origin of petroleum. Most evidence indicates that it originated from living matter in ancient seas some 500 million years ago. These species died and eventually became covered with sediments. Pressure, heat, and microbes converted what was once living matter into petroleum, which became trapped in porous rocks. It is likely that some petroleum is still being formed from sediments of dead matter. However, such a process is far too slow to consider petroleum a renewable resource.

A rock poised at the top of a hill has more potential energy—energy of position—than the same rock at the bottom of the hill.

Energy and Fossil Fuels CD-ROM WWW.

Methane is a major component of natural gas.

Fossil-fuel energy is comparable in some ways to the energy stored in a wind-up toy race car. The "winding-up" energy that was originally supplied tightened a spring in the toy. Most of that energy is stored within the spring. That stored energy is a form of **potential energy**, or energy of position. As the car moves, the spring unwinds, providing energy to the moving parts. Energy related to motion is called **kinetic energy**. Thus the movement of the car is based on converting potential energy into kinetic energy. Eventually, the toy "winds down" to a lower-energy, more stable state and stops.

In a similar manner, chemical energy, which is another form of potential energy, is stored in the bonds within chemical compounds. When an energy-releasing chemical reaction occurs, as during the burning of a fuel, bonds break and reactant atoms reorganize to form new bonds. The process yields products with different and more stable arrangements of their atoms. That is, the products have less potential energy (chemical energy) than did the original reactants. Some of the energy stored in the reactants has been released in the form of heat and light.

The combustion, or burning, of methane (CH_4) gas illustrates such an energy-releasing reaction. It can be summarized this way:

$$CH_4 \ + \ 2\,O_2 \ \longrightarrow \ CO_2 \ + \ 2\,H_2O \ + \ Energy$$

Methane gas   Oxygen gas   Carbon dioxide gas   Water

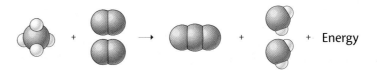

The reaction releases considerable thermal energy (heat). In fact, laboratory burner flames are based primarily on that reaction—performed, of course, under very controlled conditions. To gain a better understanding of the energy involved, imagine that the reaction takes place in two distinct steps—one involving bond-breaking and one involving bond-making.

In the first step, suppose that all the chemical bonds in one CH_4 molecule and two O_2 molecules are broken. (How many total chemical bonds need to be broken? Check the methane-burning reaction above to guide your thinking.) The result of this bond-breaking step is that separate atoms of carbon, hydrogen, and oxygen are produced. Such a bond-breaking step is an energy-requiring process, or an **endothermic** change. In an endothermic change, energy must be added to "pull apart" the atoms in each molecule. Thus energy appears as a reactant in Step 1.

An endothermic reaction can proceed only if energy is continuously supplied.

Step 1:
Energy + CH_4 + 2 O_2 $\longrightarrow$ C + 4 H + 4 O

Energy + [molecular diagrams]

To complete the methane-burning reaction, suppose that the separated atoms now join to form the new bonds needed to make the product molecules: one CO_2 molecule and two H_2O molecules. The formation of chemical bonds is an energy-releasing process, or an **exothermic** change. Because energy is given off, it appears as a product in Step 2.

Step 2:

$$C \quad + \quad 4\,H \quad + \quad 4\,O \quad \longrightarrow \quad CO_2 \quad + 2\,H_2O + Energy$$

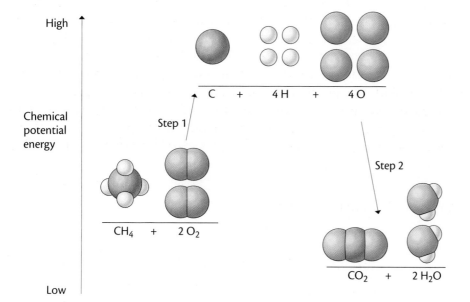

+ Energy

When methane burns, the energy released in forming carbon-oxygen bonds in CO_2 and hydrogen-oxygen bonds in H_2O is greater than the energy used to break the carbon-hydrogen bonds in CH_4 and the oxygen-oxygen bond in O_2. That is why the overall chemical change is exothermic. The complete energy summary for the burning of methane is shown in Figure 11.

Whether an overall chemical reaction is exothermic or endothermic depends on how much energy is added (endothermic process) in the bond-breaking step and how much energy is given off (exothermic process) in the bond-making step. If more energy is given off than added, the overall process is exothermic. However, if more energy is added than is given off, the overall change is endothermic.

As a general principle, if a particular chemical reaction is exothermic, then the reverse reaction is endothermic. For example, the burning of hydrogen gas—involving the formation of water from hydrogen and oxygen—is exothermic.

$$2\,H_2 + O_2 \longrightarrow 2\,H_2O + Energy$$

Therefore, the separation of water into its elements—the reverse reaction—must be endothermic.

$$Energy + 2\,H_2O \longrightarrow 2\,H_2 + O_2$$

Although exothermic reactions—which release energy—can continue on their own, it often takes the addition of "starting energy" (activation energy) to initiate the chemical change. Thus a match must be "activated" by striking it on a matchbook cover before the exothermic burning reaction starts.

Once an exothermic reaction begins, it releases energy until the reaction stops.

If a rock rolling downhill is an exothermic process, then pushing the same rock back uphill must be an endothermic process.

High

Chemical potential energy

C   +   4 H   +   4 O

Step 1

CH_4 +   2 O_2

Step 2

CO_2 +   2 H_2O

Low

Figure 11 *The formation of carbon dioxide and water from methane and oxygen gases. In Step 1, bonds are broken, which is an endothermic process. Step 2, involving bond making, releases energy, so it is an exothermic process. Because more energy is released in Step 2 than is required in Step 1, the overall reaction is exothermic.*

Oil's Well That Ends Well: The Work of a Petroleum Geologist

Susan Landon is an independent petroleum geologist. Using her knowledge of chemistry, geology, geography, biology, and mathematics, she analyzes rocks and other geologic and geographic features for indications of likely oil and natural-gas reserves.

Susan searches for the folded or faulted terrain of Earth that often contains petroleum. Hydrocarbons (oil or natural gas) may be trapped in reservoirs in these areas. Using data about a given geographic area—such as satellite images, aerial photographs, topographic maps, and information about the rocks—Susan predicts the amount of oil or natural gas that may have been generated from organic matter in the rocks, and where it might occur.

If signs are good for a substantial petroleum deposit, Susan and her team work to find investors to fund further exploration. Such exploration often includes more detailed research, analysis, and drilling. During drilling, Susan and her colleagues examine rock samples and analyze data regarding electrical and other physical properties of the rocks encountered. Based on the findings, the group recommends either completing the well or plugging it.

In Susan's opinion, a good liberal arts background with a major in geology or a related science is excellent preparation for advanced studies in geology.

Wells? Now There's a Deep Subject!

Although there are many techniques for predicting the presence of underground oil or natural gas, the only way to know for sure is to drill a well.

The process of drilling a well begins with a drilling rig consisting of a derrick—a large, vertical structure made of metal that holds and supports a drilling pipe. Rotated by a motor, the drilling pipe ends in a bit that digs and scrapes down through soil and rock. Drilling fluid (usually a special mixture of water and other materials that is called "mud") is pumped down through the pipe to cool the bit and carry chips of rock back to the surface. A geologist can examine the chips and other material brought to the surface to determine the rock type and identify the presence of oil.

Although oil or natural gas sometimes flows to the surface spontaneously as a result of underground pressure, the more common procedure is to install a pump. Pumped from the well, the oil or natural gas is transported by truck or pipeline to a refinery. At the refinery, the oil or natural gas is processed and delivered to commercial and private consumers, who convert it into everything from polypropylene socks to carpeting to household heat.

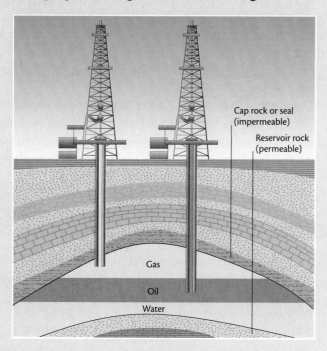

Cap rock or seal (impermeable)

Reservoir rock (permeable)

Gas

Oil

Water

> A geologist can examine the chips . . . and identify the presence of oil.

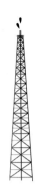

(a) Fossil fuel (potential energy)

(b) Power plant furnace (thermal energy)

(c) Generator turbine (mechanical energy)

(d) Electrical transmission lines (electrical energy)

(e) Hair dryer (mechanical, thermal, and sound energy)

Figure 12 *Tracing energy conversions from source to final use.*

B.2 ENERGY CONVERSION

Scientists and engineers have increased the usefulness of energy released from burning fuels through devices that convert thermal energy into other forms of energy. In fact, much of the energy you use daily goes through several conversions before it reaches you.

Consider the energy-conversion steps involved in the operation of a hair dryer, which is illustrated in Figure 12. What detailed, step-by-step "energy story" can you infer from the illustration? You probably noted that in the first step, stored chemical energy (potential energy) in a fossil fuel (a) is released in a power-plant furnace, producing thermal energy (heat) (b). The thermal energy then converts water in a boiler to steam that spins turbines to generate electricity. Thus the power plant converts thermal energy to mechanical energy (a form of kinetic energy) (c) and then to electrical energy (d). When the electricity reaches the hair dryer, it is converted back to thermal energy to dry your hair, and also to mechanical energy (e) as a small fan blade spins to blow the hot air. Some sound energy is also produced, as any hair-dryer user knows!

It is important to recognize that despite all of the changes involved, energy is not really consumed, or "used up," in any of these steps. Its form simply changes from chemical to thermal to mechanical to electrical, for example. This concept is summarized by the **Law of Conservation of Energy**, which states that energy is neither created nor destroyed.

WASTED ENERGY   Building Skills 4

In one sense, an automobile—whether powered by electricity, gasoline, or even solar energy—can be considered a collection of energy-converting and energy-powered devices. Imagine that you are an "energy detective," and follow some typical energy conversions in an automobile. Name the type(s) of energy involved in each step required to get a conventional automobile window defogger to operate when the car is running. See Figure 13.

Although energy-converting devices have definitely increased the usefulness of petroleum and other fuels, some of these devices have problems associated with their use. For example, potentially harmful by-products may sometimes be generated by energy-converting devices. (You will read more about this issue later in the unit.) More fundamentally, some useful energy is always "lost" when energy is converted from one form to another. That is, no energy conversion is totally efficient; some energy always becomes unavailable to do useful work.

Consider an automobile with 100 units of chemical energy stored in its fuel tank as the mixture of molecules that make up gasoline. See Figure 14 on page 202. Even a well-tuned automobile converts only about 25% of that chemical energy (potential energy) to useful mechanical energy (kinetic energy). The remaining 75% of gasoline's chemical energy is lost to the surroundings as thermal energy (heat). The following activity will allow you to see what this means in terms of gasoline consumption and expense.

ENERGY CONVERSION EFFICIENCY

Assume that your family drives 225 miles each week and that the car can travel 27.0 miles on one gallon of gasoline. How much gasoline does the car use in one year?

Questions such as this can be answered by attaching proper units to all values, then multiplying and dividing them just as though they were arithmetic expressions.

For example, the information given in the problem can be expressed this way:

$$\frac{225 \text{ miles}}{1 \text{ week}} \quad \text{and} \quad \frac{27.0 \text{ miles}}{1 \text{ gal}}$$

or, if needed, as inverted expressions:

$$\frac{1 \text{ week}}{225 \text{ miles}} \quad \text{and} \quad \frac{1 \text{ gal}}{27.0 \text{ miles}}$$

Calculating the desired answer also involves using information you already know—there are 52 weeks in one year. You also know that the desired answer must have units of gallons per year (gal/y). When all of this information is combined, and care is taken to ensure that the units are multiplied and divided to produce "gallons per year," the following expression is formed:

$$\frac{225 \text{ miles}}{1 \text{ week}} \times \frac{1 \text{ gal}}{27.0 \text{ miles}} \times \frac{52 \text{ weeks}}{1 \text{ y}} = 433 \text{ gal/y}$$

Now answer these questions.

1. Assume that an automobile that averages 27.0 miles per gallon of gasoline travels 10 500 miles each year.

   a. How many gallons of gasoline will the car burn during the year's travels?

   b. If gasoline costs $1.35 per gallon, how much would be spent on gasoline in one year?

2. Assume your automobile engine uses only 25.0% of the energy released by burning gasoline.

   a. How many gallons of gasoline are wasted each year due to energy conversion inefficiency?

   b. How much does this wasted gasoline cost at $1.35 per gallon?

3. Suppose continued research leads to a car that travels 50.0 miles on one gallon of gasoline and whose engine is 50.0% efficient.

   a. In one year, how much gasoline would be saved compared to the first car?

   b. How much money would be saved assuming a gasoline price of $1.35 per gallon?

Figure 13 *Tracing energy conversions in an automobile.*

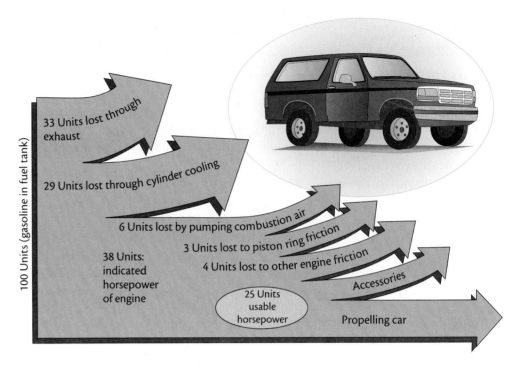

Figure 14 *Energy use in an automobile.*

Petroleum is neither limitless nor inexpensive. One way to maximize the benefits from available supplies of petroleum-based fuels is to reduce the total number of energy conversions the fuel undergoes on its way to its final use. Methods of increasing the efficiency of energy-conversion devices can also be sought.

Unfortunately, devices that convert chemical energy to heat and then to mechanical energy are typically less than 50% efficient. Solar cells, which convert solar energy to electrical energy, and fuel cells, which convert chemical energy to electrical energy, hold promise for either replacing petroleum or increasing the efficiency of its use.

Perhaps you have been wondering how the chemical energy in fuels is converted into heat and mechanical energy. What substances and chemical reactions are involved? How much energy is released? You will investigate these questions in the activity below.

B.3 COMBUSTION

Laboratory Activity

Introduction

You strike a match, and a hot, yellow flame appears. If you bring the flame close to a candlewick, the candle lights and burns. These events are so commonplace that you probably do not realize complex chemical reactions are at work.

Candle burning involves chemical reactions of the wax, which is composed of long-chain alkanes, with oxygen gas at elevated temperatures. Although many chemical reactions are involved in burning, or combustion, chemists simplify the process by usually focusing on the overall changes. For example, the burning of one component of candle wax, $C_{25}H_{52}$, can be summarized this way:

$$C_{25}H_{52}(s) \; + \; 38\,O_2(g) \; \longrightarrow \; 25\,CO_2(g) \; + \; 26\,H_2O(g) \; + \; Energy$$

Paraffin wax   Oxygen gas   Carbon   Water vapor
(Alkane)   dioxide gas

As you already know, a burning candle gives off energy—the reaction is exothermic. Thus more energy must be liberated by forming bonds in the product molecules (carbon dioxide gas and water vapor) than was originally used to break the bonds in the wax and oxygen gas reactant molecules.

Fuels provide energy as they burn. But how much energy is released? How is the quantity of released energy measured? In this activity, you will measure the heat of combustion of a candle (paraffin wax) and compare this quantity with known values for other hydrocarbons. You will also investigate relationships between the quantity of thermal energy released when a hydrocarbon burns and the structure of the hydrocarbon.

Procedure

1. Read the entire procedure and prepare a suitable data table to record all of your specified measurements—masses, volumes, and temperatures.

2. Hold a lighted match near the base of a candle so that some melted wax falls onto a 3 × 5 index card. Immediately push the base of the candle into the melted wax. Hold the candle there for a moment to fasten it to the card.

3. Determine the combined mass of the candle and index card. Record the value.

4. Carefully measure (to the nearest milliliter) about 100 mL of chilled water. (The chilled water, provided by your teacher, should be 10 to 15 °C colder than room temperature.) Pour the 100-mL sample of chilled water into an empty soft-drink can.

5. Set up the apparatus as shown in Figure 15 on page 204, but do not light the candle yet! Adjust the can so the top of the candlewick is about 2 cm from the bottom of the can.

6. Measure both room temperature and the water temperature to the nearest 0.1 °C. Record these values.

7. Place the candle under the can of water. Light the candle. As the water heats, stir it gently.

8. As the candle burns and becomes shorter, you may need to lower the can so the flame remains just below the bottom of the can. **CAUTION:** *Lower the can with great care.*

9. Continue heating until the temperature rises as far above room temperature as it was below room temperature at the start. (For example, if the water is 15 °C before heating and room temperature is 25 °C, you would heat the water to 35 °C, which is 10 °C higher than room temperature.)

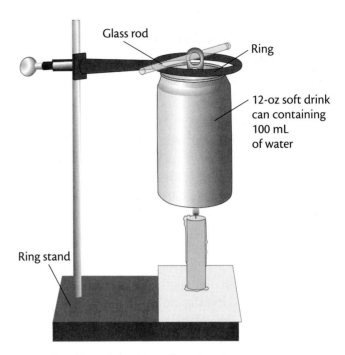

Glass rod

Ring

12-oz soft drink can containing 100 mL of water

Ring stand

Figure 15 *Apparatus for determining heat of combustion.*

10. When the desired temperature is reached, extinguish the candle flame.

11. Continue stirring the water until its temperature stops rising. Record the highest temperature reached by the water.

12. Determine the mass of the cooled candle and index card, including all wax drippings.

13. Wash your hands thoroughly before leaving the laboratory.

Calculations

A characteristic property of a substance is the quantity of heat needed to raise the temperature of one gram of the substance by 1 °C. This value is called the **specific heat capacity** of the substance. The specific heat capacity of liquid water is about 4.2 J/(g·°C) (joules per gram per °C). This means that for each degree Celsius that liquid water is heated, each gram of the water absorbs 4.2 J of thermal energy.

Suppose a 10.0-g water sample is heated from 25.0 °C to 30.0 °C, a temperature increase of 5.0 °C. How much thermal energy must have been added to the water?

The answer can be reasoned this way. It takes 4.2 J to raise the temperature of 1 g water by 1 °C. In this example, however, there is 10 times more water and a temperature increase that is 5 times greater. Thus 10.0 × 5.0, or 50 times more thermal energy is needed. So, the specific heat must be multiplied by 50 to obtain the answer: 50 × 4.2 J = 210 J. It takes 210 J to increase the temperature of 10.0 g water by 5.0 °C.

The quantity of thermal energy given off when a certain amount of a substance burns is called the **heat of combustion**. The heat of combustion can be expressed as the thermal energy released when either one gram of

Specific heat capacity is sometimes shortened to the term "specific heat."

The exact specific heat capacity of water is 4.184 J/g. However, the rounded-off value of 4.2 J/g is adequate for and equally useful in this laboratory activity.

substance burns or one mole of substance burns. If the amount of substance being burned is one mole, the quantity of thermal energy is called the **molar heat of combustion**. Using your laboratory data, you can calculate the heat of combustion of paraffin wax.

1. Calculate the mass of water heated. (*Hint:* The density of liquid water is 1.0 g/mL. Thus each milliliter of water has a mass of 1.0 g.)

2. Calculate the total rise in temperature of the water.

3. Calculate how much thermal energy was used to heat the water sample. Use values from the two preceding steps to reason out the answer, as illustrated in the sample problem.

4. Calculate the total mass of paraffin wax burned.

5. Calculate the heat of combustion of paraffin, expressed

   a. in units of joules per gram (J/g) of paraffin.
   b. in units of kJ/g.

   Hint: Assume that all the energy released by the burning paraffin wax is absorbed by the water. Divide the total thermal energy by the mass of paraffin burned to get the answer.

$$\text{Heat of combustion} = \frac{\text{thermal energy released (Step 3)}}{\text{mass of paraffin burned (Step 4)}}$$

(1 kJ = 1000 J)

Questions

Your teacher will collect your heat of combustion data, expressed in units of kilojoules per gram (kJ/g) of paraffin. Use the combined results of your class to determine a "best" estimate for this value. Will you decide to use the average value of the class results or the median value? Why? Use your selected value to answer the following questions.

1. How does your experimental heat of combustion (in kJ/g) for paraffin wax, $C_{25}H_{52}$, compare to the accepted heat of combustion for propane, C_3H_8? See Figure 16.

Heats of Combustion			
Hydrocarbon	Formula	Heat of Combustion (kJ/g)	Molar Heat of Combustion (kJ/mol)
Methane	CH_4	55.6	890
Ethane	C_2H_6	52.0	1560
Propane	C_3H_8	50.0	2200
Butane	C_4H_{10}	49.3	2859
Pentane	C_5H_{12}	48.8	3510
Hexane	C_6H_{14}	48.2	4141
Heptane	C_7H_{16}	48.2	4817
Octane	C_8H_{18}	47.8	5450

Figure 16 *Hydrocarbon heats of combustion.*

2. How do the molar heats of combustion (in kJ/mol) for paraffin and propane compare? (*Hint:* To make this comparison, first calculate the thermal energy released when one mole of paraffin burns. Because one mole of paraffin $[C_{25}H_{52}]$ has a mass of 352 g, the molar heat of combustion will be 352 times greater than the heat of combustion expressed as kJ/g.)

3. Explain any differences noted between paraffin and propane in your answers to Questions 1 and 2. (*Hint:* Keep in mind the calculation you completed in Question 2.)

4. In your view, which hydrocarbon—paraffin or propane—is the better fuel? Explain your answer.

5. In calculating heats of combustion, you assumed that all thermal energy from the burning fuel went to heating the water.
   a. Is this a good assumption? Explain.
   b. What other laboratory conditions or assumptions might cause errors in your calculated values?

B.4 USING HEATS OF COMBUSTION

Energy from Combustion

With abundant oxygen gas and complete combustion, the burning of a hydrocarbon can be described by the equation

Hydrocarbon + Oxygen gas $\longrightarrow$ Carbon dioxide + Water + Thermal energy

Energy is written as a product of the reaction because energy is released when a hydrocarbon burns. The combustion of a hydrocarbon is a highly exothermic reaction.

The equation for burning ethane (C_2H_6) is

$$2\,C_2H_6 + 7\,O_2 \longrightarrow 4\,CO_2 + 6\,H_2O + \textbf{?}\ kJ\ \text{thermal energy}$$

To complete this equation, the correct quantity of thermal energy involved must be included. Figure 16 indicates that ethane's molar heat of combustion is 1560 kJ/mol. That is, burning one mole of ethane releases 1560 kilojoules of energy. But according to the chemical equation above, two moles of ethane $(2\,C_2H_6)$ are burned. Thermal energy must be "balanced" in terms of all other reactants and products. Thus the total thermal energy released will be twice that released when one mole of ethane burns:

$$2\ \cancel{mol} \times \frac{1560\ kJ}{1\ \cancel{mol}} = 3120\ kJ$$

The complete combustion equation for ethane is

$$2\,C_2H_6 + 7\,O_2 \longrightarrow 4\,CO_2 + 6\,H_2O + 3120\ kJ$$

As you found out in the preceding laboratory activity (and as is indicated in Figure 16), heats of combustion can also be expressed as energy produced when one gram of hydrocarbon burns (kJ/g). That information is very useful in determining how much energy is released when a certain mass of fuel is burned.

For example, how much thermal energy would be produced by burning 12.0 g octane, C_8H_{18}? Figure 16 indicates that the burning of 1.00 g octane releases 47.8 kJ. Burning 12.0 times more octane will produce 12.0 times more thermal energy, or

$$12.0 \times 47.8 \text{ kJ} = 574 \text{ kJ}.$$

The calculation can also be written this way:

$$12.0 \text{ g octane} \times \frac{47.8 \text{ kJ}}{1 \text{ g octane}} = 574 \text{ kJ}$$

HEATS OF COMBUSTION

Building Skills 6

This activity will give you a better understanding of the energy involved in burning hydrocarbon fuels. Use Figure 16 (page 205) to answer the following questions. The first one is worked out as an example.

1. How much energy (in kilojoules) is released by completely burning 25.0 mol hexane, C_6H_{14}?

According to Figure 16, the molar heat of combustion of hexane is 4141 kJ. This means that 4141 kJ of energy is released when 1.00 mol hexane burns. So burning 25.0 times more fuel will produce 25.0 times more energy:

$$25.0 \text{ mol } C_6H_{14} \times \frac{4141 \text{ kJ}}{1 \text{ mol } C_6H_{14}} = 104\,000 \text{ kJ}$$

Burning 25.0 mol hexane would thus release 104 000 kJ of thermal energy.

2. Write a chemical equation that includes thermal energy for the complete combustion of these alkanes:

   a. propane
   b. butane

3. Examine the data summarized in Figure 16.

   a. How does the trend in heats of combustion for hydrocarbons expressed as kJ/g compare with the trend expressed as kJ/mol?
   b. Assuming the trend applies to larger hydrocarbons, predict the heat of combustion for decane, $C_{10}H_{22}$, expressed as kJ/g decane and kJ/mol decane.
   c. Which prediction in Question 3b was easier? Why?

4. a. How much thermal energy is produced by burning two moles of octane?
   b. How much thermal energy is produced by burning one gallon of octane? (A gallon of octane has a volume of about 3.8 liters. The density of octane is 0.70 g/mL.)
   c. Suppose a car operates so inefficiently that only 16% of the thermal energy from burning fuel is converted to useful "wheel-turning" (mechanical) energy. How many kilojoules of useful energy would be stored in a 20.0-gallon tank of gasoline? (Assume that octane burning and gasoline burning produce the same results.)

5. The molar heat of combustion of carbon contained in coal is 394 kJ/mol C.

   a. Write a chemical equation for burning the carbon contained in coal. Include the thermal energy produced.
   b. Gram for gram, which is the better fuel—carbon or octane? Explain your answer using calculations.
   c. In what applications might coal replace petroleum-based fuel?
   d. Describe one application in which coal would be a poor substitute for petroleum.

As automobile use grows around the world, demand for gasoline continues to increase rapidly. Because the gasoline fraction in a barrel of crude oil normally represents only about 18% of the total, researchers have been anxious to find a way to increase this yield. One promising method has been based on the discovery that it is possible to alter the structures of some petroleum hydrocarbons so that 47% of a barrel of crude oil can be converted to gasoline. This important chemical technique deserves further attention.

B.5 ALTERING FUELS

As you might expect, not all fractions of hydrocarbons obtained from petroleum are in equal demand or use at any particular time. The market for one petroleum fraction may be much less profitable than the market for another. For example, the invention of electric lightbulbs caused a rapid decline in the use of kerosene lanterns in the early 1900s. As a result, the kerosene fraction of petroleum, composed of hydrocarbon molecules with 12 to 16 carbon atoms, became a surplus commodity. On the other hand, automobiles dramatically increased the demand for the gasoline fraction (C_5 to C_{12}) that had earlier been discarded. The gasoline fraction could not be used safely in lanterns—it exploded!

Chemists and chemical engineers are adept at modifying or altering available chemical resources to meet new needs. Such alterations might involve converting less-useful materials to more-useful products, or—as in the case of kerosene in the early 1900s—converting a low-demand material into high-demand materials.

Cracking

Recall that the gasoline fraction obtained from crude oil refining includes hydrocarbons with 5 to 12 carbons per molecule.

By 1913, chemists had devised a process for converting larger molecules in kerosene into smaller, gasoline-sized molecules by heating the kerosene to 600 to 700 °C. The process of converting large hydrocarbon molecules into smaller ones through the application of heat and a catalyst is known as **cracking**. Through cracking, a 16-carbon molecule, for example, might produce two 8-carbon molecules:

$$C_{16}H_{34} \longrightarrow C_8H_{18} + C_8H_{16}$$

In practice, molecules with up to about 14 carbon atoms can be produced through cracking. Molecules with 5 to 12 carbon atoms are particularly useful in gasoline, which remains the most important commercial product of refining. Some C_1 to C_4 molecules (C_1, C_2, C_3, and C_4) produced in cracking are immediately burned, keeping the temperature high enough for more cracking to occur.

Today, more than a third of all crude oil undergoes cracking. The process has been improved by adding catalysts such as aluminosilicates. A **catalyst** increases the speed of a chemical reaction but is not itself used up. Catalytic cracking is more energy efficient because it occurs at a lower temperature—500 °C rather than 700 °C.

Gasoline is composed mainly of straight-chain alkanes, such as hexane (C_6H_{14}), heptane (C_7H_{16}), and octane (C_8H_{18}). As a result, it burns very rapidly. The rapid burning causes engine "pinging," or "knocking," and may contribute to engine problems. Branched-chain alkanes burn more satisfactorily in automobile engines; they do not ping as much. The structural isomer of octane shown below has excellent combustion properties in automobile engines. This octane isomer is known chemically as 2,2,4-trimethylpentane. Can you see how the name of the molecule is related to its structure? For convenience, this compound is frequently referred to by its common name, isooctane.

$$CH_3-\underset{\underset{CH_3}{|}}{\overset{\overset{CH_3}{|}}{C}}-CH_2-\underset{\underset{CH_3}{|}}{CH}-CH_3$$

Isooctane, C_8H_{18}

Octane Rating

As you probably know, gasoline is sold in a variety of grades—and at corresponding prices. A common reference standard for gasoline quality is the octane scale. On this scale, isooctane, the branched-chain hydrocarbon you just learned about, is assigned an **octane number** of 100. Straight-chain heptane (C_7H_{16}), a fuel with very poor engine performance, is assigned an octane number of zero. Gasoline samples can be rated in comparison with isooctane and heptane. For example, a gasoline with knocking characteristics similar to a mixture of 87% isooctane and 13% heptane has an octane number of 87. The higher the octane number of a gasoline, the better its antiknock characteristics. Octane ratings in the high 80s and low 90s (85, 87, 92) are quite common, as a survey of nearby gas pumps, like those pictured in Figure 17 on page 210, will reveal.

Assigning an octane number of 100 to isooctane is arbitrary and does not mean that it has the highest possible octane number. In fact, several fuels burn more efficiently in engines than isooctane does and have octane numbers above 100.

Prior to the mid-1970s, the octane rating of gasoline was increased at low cost by adding a substance such as tetraethyl lead, $(C_2H_5)_4Pb$, to the fuel. This additive slowed down the burning of straight-chain gasoline molecules and added about three points to "leaded" fuel's octane rating. Unfortunately, lead from the treated gasoline was discharged into the atmosphere along with other vehicle exhaust products. Lead is harmful to the environment and, as a result, such lead-based gasoline additives are no longer used.

Oxygenated Fuels

The phaseout of lead-based gasoline additives meant that alternative octane-boosting supplements were required. A group of additives called **oxygenated fuels** have been blended with gasoline to enhance its octane rating. The molecules of these additives contain oxygen in addition to carbon and hydrogen.

Although oxygenated fuels actually deliver less energy per gallon than regular gasoline hydrocarbons do, their economic appeal stems from their ability to increase the octane number of gasoline while reducing exhaust-gas pollutants. In particular, oxygenated fuels encourage more complete combustion, producing lower emissions of air pollutants such as carbon monoxide (CO).

A common oxygenated fuel is methanol (methyl alcohol, CH_3OH), which is added to gasoline at distribution locations. In addition to its octane-boosting properties, methanol can be made from natural gas, coal, corn, or wood—a contribution toward conserving nonrenewable petroleum resources. A blend of 10% ethanol (ethyl alcohol, CH_3CH_2OH) and 90% gasoline, sometimes called **gasohol**, can be used as an oxygenated fuel in nearly all modern automobiles without associated engine adjustments or problems.

Figure 17 *Octane ratings are posted on gasoline pumps.*

Methyl tertiary-butyl ether, MTBE, which has an octane rating of 116, was initially introduced in the late 1970s as an octane-boosting fuel additive. In the 1990s, the role of MTBE as a pollution-reducing oxygenated fuel became increasingly important. In fact, MTBE became the most common oxygenated fuel additive in gasoline. At its peak, annual U.S. production of MTBE increased to more than 4 billion gallons (about 16 gallons per person), making it one of the top ten industrial chemicals.

By the late 1990s, however, evidence of contamination of groundwater and drinking water supplies due to seepage of MTBE from defective underground gasoline storage systems began to mount. MTBE dissolves readily in water, is resistant to microbial decomposition, and is difficult to remove in water-treatment processes. The unpleasant taste and odor that MTBE imparts to water, even at concentrations below those regarded as a public health concern, triggered consumer complaints. So although MTBE has been credited by the U.S. Environmental Protection Agency for substantial reductions in emissions of air pollutants from gasoline-powered vehicles in the 1990s, its reduced use or complete elimination as an oxygenated fuel is under active consideration, research, and policy debate.

A promising new fuel additive is MTHF, methyltetrahydrofuran. MTHF has an octane rating of 87, equal to that of regular unleaded gasoline, and the ability to increase the oxygenated level of the fuel. An added advantage of using MTHF is that it can be obtained from renewable resources, such as paper-mill waste products.

$$CH_3 - O - \overset{\overset{\displaystyle CH_3}{|}}{\underset{\underset{\displaystyle CH_3}{|}}{C}} - CH_3$$

Methyl tertiary-butyl ether (MTBE).

A BURNING ISSUE   ChemQuandary 2

Methanol (CH_3OH) and ethanol (CH_3CH_2OH) are used as gasoline additives or substitutes. Their heats of combustion are 23 kJ/g and 30 kJ/g, respectively.

Consider the chemical formulas of methanol and ethanol. Gram for gram, why are heats of combustion of methanol and ethanol considerably less than those of any hydrocarbons discussed so far? See Figure 16, page 205.

Other octane-boosting strategies involve altering the structures of hydrocarbon molecules in petroleum. This works because branched-chain hydrocarbons burn more satisfactorily than straight-chain hydrocarbons. (Recall isooctane's octane number compared with that of heptane.) Straight-chain hydrocarbons are converted to branched-chain hydrocarbons by a process called isomerization. During isomerization, hydrocarbon vapor is heated with a catalyst:

$$CH_3 - CH_2 - CH_2 - CH_2 - CH_2 - CH_3 \quad \xrightarrow[\text{Catalyst}]{\text{Heat}} \quad CH_3 - \overset{\overset{\displaystyle CH_3}{|}}{CH} - \overset{\overset{}{}}{\underset{\underset{\displaystyle CH_3}{|}}{CH}} - CH_3$$

$C_6H_{14}(g)$ $C_6H_{14}(g)$
Straight-chain   Branched-chain
isomer   isomer

The branched-chain alkanes produced are blended with the C_5 to C_{12} molecules obtained from cracking and distillation, producing a high-quality gasoline.

Although cracked and isomerized molecules improve the burning of gasoline, they also increase its cost. One reason for this increase is the extra fuel used in manufacturing such gasoline.

B.6 FUEL MOLECULES IN TRANSPORTATION

Making Decisions

Now that you have examined the use of petroleum as a fuel, it is time to revisit the TV advertisement for the ARL-600 vehicle that opened this unit. First, use what you have learned about petroleum to answer the questions below. Then, write one or two paragraphs that critically evaluate the claims made in that advertisement.

1. In what ways does the manufacture of an ARL-600 use petroleum as a fuel?

2. In what ways does the operation of an ARL-600 use petroleum as a fuel?

3. What viable alternatives to petroleum can you suggest for fuel uses listed in Questions 1 and 2?

4. What direct, nonfuel uses of petroleum or its fractions are involved in the operation of an ARL-600?

5. What viable alternatives to petroleum can you suggest for the direct uses listed in Question 4?

6. What facts or concepts involving "energy from petroleum" in the ARL-600 advertisement are misleading? Why?

7. How would you correct the facts or concepts identified in Question 6?

Questions and Answers

Section C will expand your knowledge of the "building" role of petroleum, focusing on it as a source of substances from which an impressive array of useful compounds and materials can be produced. This additional information will help you further decide about the accuracy of claims found in the new-car advertisement.

SECTION SUMMARY

Reviewing the Concepts

♦ **The energy released by burning fossil fuels such as petroleum, natural gas, or coal originates from solar energy originally captured by prehistoric green plants during photosynthesis.**

1. From a chemical viewpoint, why is petroleum sometimes considered "buried sunshine"?

2. Why is the energy stored in petroleum more useful than solar energy for most applications? Give at least three reasons in your explanation.

3. Write a word equation for photosynthesis. Include energy in your equation.

4. Compare the equation for the burning of hydrocarbons to the equation for photosynthesis. How are they related to each other?

5. How were prehistoric green plants converted into today's petroleum reserves?

♦ **Chemical energy, which is a form of potential energy, is stored in the bonds within chemical compounds. In all chemical reactions, bonds break and atoms rearrange to form new bonds.**

6. Classify each of the following as an example of kinetic or potential energy.

   a. a skateboard at the top of a hill
   b. a charged battery
   c. a rolling bowling ball
   d. a gallon of gasoline
   e. a waterfall

7. Based on its structural formula, does a molecule of methane or a molecule of butane have more potential energy? Explain your answer.

♦ **The energy change in a chemical reaction equals the difference between the energy required to break reactant bonds and the energy released in forming product bonds. If more energy is released than added, the reaction is exothermic. If more energy is added than released, the reaction is endothermic.**

8. The burning of a candle is an exothermic reaction. Explain this fact in terms of the quantity of energy stored in the bonds of the reactants compared with the quantity of energy stored in the bonds of the products.

9. Using Figure 11 on page 197 as a model, illustrate the energy change when hydrogen gas reacts with oxygen gas to produce thermal energy and water.

10. For each of the following reactions, determine whether bond-breaking or bond-making involves more energy.

    a. burning wood in a campfire
    b. activating a chemical "cold pack"
    c. burning oil in a lamp

- ◆ Thermal energy can be converted into other forms of energy, such as mechanical, electrical, or potential energy. Although energy can neither be created nor destroyed, useful energy is lost each time energy is converted from one form to another.

11. Identify the energy conversions involved in making each of the following events possible.

   a. riding a bike
   b. illuminating a lightbulb
   c. producing electricity in a wind-powered generator
   d. running a gasoline-powered lawn mower

12. In powering an automobile, 25% of the energy is said to be useful. What happens to the other 75% of the energy?

13. Explain what is meant by energy conversion efficiency.

14. One gallon of gasoline produces about 132 000 kJ of energy when burned. Assume that an automobile is 25% efficient in converting this energy into useful work.

   a. How much energy is "wasted" when a gallon of gasoline burns?
   b. What happens to this "wasted" energy?

- ◆ When a hydrocarbon burns completely, it reacts with oxygen gas in the air to produce carbon dioxide and water vapor.

15. Write a balanced chemical equation for the complete combustion of

   a. methane.
   b. propane.
   c. octane.

16. When candle wax (a mixture of hydrocarbons) burns, it seems to disappear. What actually happens to the wax?

17. a. What are the products of the complete burning of octane?
    b. What are the products of the complete burning of a mixture of C_8H_{18} and C_7H_{16}?
    c. Compare and comment on your answers to Questions 17a and 17b.

- ◆ The quantity of thermal energy (measured in joules) required to raise the temperature of one gram of a material by 1 °C is its specific heat capacity. The molar heat of combustion is the quantity of energy released when one mole of a substance burns.

18. Water gas (a 50-50 mixture of CO and H_2) is made by the reaction of coal with steam. Because the United States has substantial coal reserves, water gas might serve as a substitute fuel for natural gas (composed mainly of methane, CH_4). Water gas burns according to this equation:

$$CO + H_2 + O_2 \longrightarrow CO_2 + H_2O + 525 \text{ kJ}$$

   a. How does water gas compare to methane in terms of thermal energy produced?
   b. If a water gas mixture containing 10 mol CO and 10 mol H_2 were completely burned in O_2, how much thermal energy would be produced?

19. How much energy is released when a 5600-g sample of water cools from 99 °C to 28 °C? The specific heat capacity of water is 4.18 J/(g·°C).

20. The combustion of acetylene, C_2H_2 (used in a welder's torch), can be represented as:

$$2 C_2H_2 + 5 O_2 \longrightarrow 4 CO_2 + 2 H_2O + 2512 \text{ kJ}$$

   a. What is the molar heat of combustion of acetylene in kilojoules per mole?
   b. If 12 mol acetylene burns fully, how much thermal energy will be produced?

21. a. List two factors that would help you decide which hydrocarbon fuel to use in a particular application.
    b. Explain the importance of each factor.

22. In a laboratory activity, a student team measures the heat released by burning heptane (C_7H_{16}). Using the following data, calculate the molar heat of combustion of heptane in kJ/mol.

Mass of water	179.2 g
Initial water temperature	11.6 °C
Final water temperature	46.1 °C
Mass of heptane burned	0.585 g

- Larger molecules in petroleum's kerosene fraction (C_{12} to C_{16}) can be converted to smaller molecules found in the gasoline fraction (C_5 to C_{12}). The process, called cracking, is speeded up by use of a catalyst.

23. List some molecules that might be formed when the following substances are cracked.

    a. $C_{16}H_{34}$
    b. $C_{18}H_{38}$

24. Why are catalysts used during the cracking process?

25. Explain why only small amounts of catalysts are needed to crack large amounts of petroleum.

- The octane scale provides a measure of a fuel's burning performance in an engine. Additives may increase gasoline's burning performance.

26. Explain the meaning of a fuel's octane rating.

27. A premium gasoline has an octane rating of 93. To what isooctane-heptane mixture does this rating compare?

28. List two ways to increase a fuel's octane rating.

29. Compare the molecular structures of octane and isooctane.

30. How does the addition of oxygenating compounds affect a fuel's octane rating?

31. Why is tetraethyl lead no longer used as a gasoline additive?

Connecting the Concepts

32. In a laboratory activity, a student burns 4.2 g ethanol (C_2H_5OH). The molar heat of combustion of ethanol is 1366 kJ/mol.

    a. How much energy is released?
    b. The heat from this reaction is used to warm a 468-g sample of water in a calorimeter. The water temperature changes from 21 °C to 89 °C. How much energy is absorbed by the water? The specific heat capacity of water is 4.18 J/(g·°C).

    c. Compare the quantity of heat released by the reaction to the quantity of heat absorbed by the calorimeter. What accounts for the difference?

33. Was the formation of petroleum the result of chemical or physical changes? Explain your answer.

34. You are given a choice of three fuels for use in heating your home—candles, butane, or gasoline. Based on their heats of combustion, which one would you choose? Why?

Extending the Concepts

35. Powdered aluminum or magnesium metal releases considerable heat when burned. (The heats of combustion of aluminum and magnesium are 31 kJ/g and 25 kJ/g, respectively.) Would these highly energetic powdered metals be good fuel substitutes for petroleum? Explain.

36. One theory for the extinction of the dinosaurs is that Earth was struck by a meteorite. The enormous cloud of dust and debris that resulted blotted out sunlight for a period of several years. How does this sequence of events explain the extinction of the dinosaurs?

PETROLEUM AS A BUILDING SOURCE

Just as an architect uses available construction materials to design a building, a chemist—a "molecular architect"—uses available molecules to design new molecules. Architects must know about the structures and properties of common construction materials. Likewise, chemists must understand the structures and properties of their raw materials—the "builder molecules." In this section, you will explore the structures of common hydrocarbon builder molecules and some materials made from them.

C.1 CREATING NEW OPTIONS: PETROCHEMICALS

Until the early 1800s, all objects and materials used by humans were either created directly from what are called found materials, such as wood or stone, or crafted from metals, glass, and clays. Available fibers included cotton, wool, linen, and silk. All medicines and food additives came from natural sources. The only plastics were those made from wood (celluloid) and animal materials (shellac).

Today many common objects and materials created by the chemical industry are unlike anything seen or used by citizens of the 1800s or even the mid-1900s. Compounds produced from oil or natural gas are called **petrochemicals**. Some petrochemicals, such as detergents, pesticides, pharmaceuticals, and cosmetics, are used directly. Most petrochemicals, however, serve as raw materials in the production of other synthetic substances, particularly a wide range of plastics.

Plastics include paints, fabrics, rubber, insulation materials, foams, adhesives, molding, and structural materials. Worldwide production of petroleum-based plastics is more than four times that of aluminum products.

The astounding fact is that it takes relatively few builder molecules (small-molecule compounds) to make thousands of new substances. One particularly important builder molecule is ethene, C_2H_4, a hydrocarbon compound commonly called ethylene. The structural formula for ethene can be seen in the equation shown below. Because of the high reactivity of its double bond, ethene is readily transformed into many useful products. A simple example—the formation of ethanol (ethyl alcohol) from water and ethene—illustrates how ethene reacts.

> Double bonds will be discussed in more detail on page 220.

$$
\begin{array}{ccccc}
\underset{\text{Ethene}}{\overset{\displaystyle H \diagdown \quad \diagup H}{\underset{\displaystyle H \diagup \quad \diagdown H}{C=C}}} & + & \underset{\text{Water}}{H-OH} & \xrightarrow{\text{Acid Catalyst}} & \underset{\text{Ethanol}}{\overset{\displaystyle H \ H}{\underset{\displaystyle H \ OH}{H-C-C-H}}}
\end{array}
$$

In this reaction, the water molecule "adds" to the double-bonded carbon atoms by placing an H— on one carbon and an —OH group on the other. This type of chemical change is called an **addition reaction**.

Ethene can also undergo an addition reaction with itself. Because the added ethene molecule contains a double bond, another ethene molecule can be added, and so on. This creates a long-chain substance called polyethene, commonly known as polyethylene, a **polymer**. A polymer is a large molecule typically composed of 500 to 20 000 or more repeating units called **monomers**. In polyethene the repeating unit is an ethene (ethylene) monomer. The chemical reaction that produces polyethene can be written this way:

| Ethene (Ethylene) | The growing polymer chain | Polyethene (Polyethylene) |

Polymers formed in reactions such as this are called—sensibly enough—**addition polymers**. Polyethene, commonly used in grocery bags and packaging, is one of the most important addition polymers. The United States produces millions of kilograms of polyethene every year.

A great variety of addition polymers can be made from monomers that closely resemble ethene. The most common variation is to replace one or more of ethene's hydrogen atoms with an atom or atoms of another element. In the following examples, note the atom or atoms that replace hydrogen in each of these monomers and polymers.

Vinyl chloride   Polyvinyl chloride

Acrylonitrile   Polyacrylonitrile

Styrene   Polystyrene

The atoms that compose the monomers dictate the properties of the polymer. Thus polyethene and polyvinyl chloride are fundamentally different materials. However, as with many modern materials, the properties of a polymer are often altered to meet a variety of needs and to produce a multitude of products.

Perhaps you are wondering why two names are shown for the same substances—ethene/ethylene and polyethene/polyethylene. Chemists worldwide have developed a system for naming chemical substances that makes it easy for them to communicate with one another. However, some substances were given names long before this system was universally accepted. Some "common names" of substances, such as ethylene and polyethylene, continue in wide use.

The hexagon-shaped ring depicted in styrene and polystyrene represents benzene, a substance that will be discussed on page 223.

Figure 18: *Products commonly made from (a) polyvinyl chloride, (b) polyacrylonitrile, and (c) polystyrene.*

(a)

(b)

(c)

POLYMER STRUCTURE AND PROPERTIES

UNMODIFIED, THE ARRANGEMENT of covalent bonds in long, stringlike polymer molecules causes the molecules to coil loosely. A collection of polymer molecules (such as those in a piece of molten plastic) can intertwine, much like strands of cooked spaghetti. In this form the polymer is flexible and soft.

1. Using a pencil line on paper to represent a linear polymer, draw a collection of loosely coiled polymer molecules.

For most polymers like the ones you just drew, flexibility and ductility depend upon temperature. Ductility refers to the ability of a material to be drawn out, as into wires. When the material is warm, the polymer chains can slide past one another easily. However, the polymer becomes more rigid when it cools. Such polymers, called thermoplastics, are found in many everyday products.

The flexibility of a polymer can also be enhanced by adding molecules that act as internal lubricants between the polymer chains. For example, untreated polyvinyl chloride (PVC) is used in rigid pipes and house siding. With added lubricant, polyvinyl chloride becomes flexible enough to be used in raincoats and inflatable pool toys.

The reactions that form polymer chains can also take place perpendicular to the main chain, forming side chains. These polymers are called **branched polymers**. The branched form of polyethylene is shown below. The degree of branching can be controlled by adjusting reaction conditions.

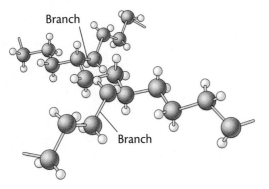

2. Draw at least two different models of branched chain polymers. Try to vary the representations—the forms of branched polymers can differ greatly.

Branching changes the properties of a polymer by affecting the ability of chains to slide past one another and by altering intermolecular forces.

Another way to alter the properties of polymers is through cross-linking. Polymer rigidity can be increased if the polymer chains are cross-linked so that they can no longer move or slide readily. You can see this for yourself if you compare the flexibility of a rubber band with that of a tire tread. Polymer cross-linking is much greater in the tire.

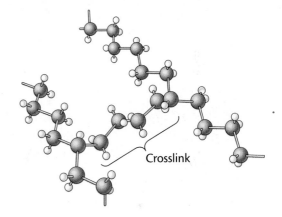

Above is cross-linked form of polyethylene.

3. Draw several linear polymer chains that have been cross-linked.

4. Draw several cross-linked, branched polymers. Then explain how cross-linking branched polymers compares with cross-linking linear polymers.

Polymer strength and toughness can also be controlled. To do this, polymer chains are arranged so that they lie in the same general direction. The aligned chains are then stretched until they uncoil. Polymers remaining uncoiled after this treatment make strong, tough films and fibers. Such materials include polyethene, used in everything from plastic grocery bags to artificial ice rinks, and polyacrylonitrile, found in fabrics.

5. Draw several aligned linear polymer chains.

6. Draw several aligned and cross-linked linear polymer chains.

C.2 BEYOND ALKANES

Carbon is a versatile building-block atom. It can form bonds with other atoms in several different ways. As you learned in Section A, in alkane molecules each carbon atom is bonded to four other atoms. Compounds such as alkanes are called saturated hydrocarbons because each carbon atom forms as many single bonds as it can by bonding to four other atoms. In some hydrocarbon molecules, however, carbon atoms bond to three other atoms, not four. This series of hydrocarbons is called the **alkenes**. The simplest member of the alkenes was briefly introduced earlier in this section—ethene, C_2H_4.

The carbon-carbon bonding that characterizes the alkenes is a **double covalent bond**. In a double covalent bond, four electrons are shared between the bonding partners. Compounds containing carbon-carbon double bonds are described as **unsaturated**—not all carbon atoms are bonded to their full capacity with four other atoms. Because of their double bonds, alkenes are more chemically reactive—and therefore better builder molecules—than are alkanes.

The substituted alkenes make up another class of important builder molecules. In addition to carbon and hydrogen, these molecules contain one or more other elements, such as oxygen, nitrogen, chlorine, or sulfur. Adding atoms of other elements to hydrocarbon structures significantly changes their chemical reactivity.

Even molecules composed of the same elements can have quite different properties. For example, a molecule of ethanol (C_2H_6O)—also called grain alcohol—and a molecule of ethylene glycol ($C_2H_6O_2$)—often used as antifreeze—both contain carbon, hydrogen, and oxygen. The dramatic differences in their properties and uses result from the different arrangement of atoms in the two molecules.

C.3 THE BUILDERS   Laboratory Activity

Introduction

This activity, in which you will use models to simulate various arrangements of atoms, will help you to become more familiar with the alkenes and their polymers.

Procedure

Part 1: Alkenes

1. Examine the electron-dot and structural formulas for ethene, C_2H_4. Confirm that each atom has attained a filled outer shell of electrons.

$$H:C::C:H$$

Electron-dot
formula

$$H-C=C-H$$

Structural
formula

CH_2CH_2 or C_2H_4

Molecular formula

The names of the alkenes follow a pattern much like that of the alkanes. The first three alkenes are ethene, propene, and butene. (Why is it impossible to have methene, a one-carbon alkene?) The same prefixes that you learned for alkanes are used to indicate the number of carbon atoms in the molecule's longest carbon chain. However, each alkene name ends in *-ene* instead of *-ane*.

2. Recall that the alkane general formula is C_nH_{2n+2}. Examine the molecular formulas of ethene (C_2H_4) and butene (C_4H_8). What general formula for alkenes do the molecular formulas suggest?

3. Assemble a model of an ethene molecule and a model of an ethane (C_2H_6) molecule. Compare the arrangements of atoms in the two models. Rotate the two carbon atoms in ethane about the single bond. Then try a similar rotation with ethene. What do you observe? Can you build a molecule in which you can perform a rotation about a double bond? Write a general rule to summarize your findings.

4. Build a model of butene (C_4H_8). Compare your model to those made by others. Remember that alkenes must contain a double bond.

   a. How many different arrangements of atoms in a C_4H_8 chain appear possible? Each arrangement represents a different substance—another example of isomers!

   b. Which structural formulas in Figure 19 correspond to models built by you or your classmates?

As with alkanes, alkenes are named according to the length of their longest carbon chain. The carbon atoms are numbered, beginning at the end of the chain closest to the double bond. The name of each isomer starts with the number assigned to the first double-bonded carbon atom. Look again at the butene isomer structures in Figure 19 and confirm this naming convention with 1-butene and 2-butene.

5. Does each of the following pairs represent isomers, or are they the same substance?

   a. $CH_2{=}CH{-}CH_2{-}CH_3$ or $CH_3{-}CH_2{-}CH{=}CH_2$

   b. $CH_2{=}\underset{\underset{CH_3}{|}}{C}{-}CH_3$ or $CH_3{-}\underset{\overset{||}{CH_2}}{C}{-}CH_3$

6. How many isomers of propene (C_3H_6) are there? Support your answer with the appropriate structure(s).

7. Are these two structures isomers or the same substance? Explain.

$$\begin{array}{c} CH_2{-}CH_2 \\ | \qquad\quad | \\ CH_2 \quad CH_2 \\ \diagdown\, CH_2\,\diagup \end{array} \qquad \begin{array}{c} CH_3{-}CH{-}CH_2 \\ | \qquad\quad | \\ CH_2{-}CH_2 \end{array}$$

8. Based on your knowledge of molecules with single and double bonds between carbon atoms, assemble a model of a hydrocarbon molecule with a triple bond. Your completed model represents a member of the hydrocarbon series known as **alkynes**. Based on your understanding of how alkanes and alkenes are named, write structural formulas for

   a. ethyne, commonly called acetylene.

   b. 2-butyne.

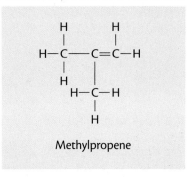

1-Butene, or simply butene

2-Butene

Methylpropene

Figure 19: *Three isomers of butene, C_4H_8.*

9. Are alkynes saturated or unsaturated hydrocarbons? Explain.

Part 2: Compounds of Carbon, Hydrogen, and Singly-Bonded Oxygen

10. Assemble as many different molecular models as possible using all nine of these atoms:

    2 carbon atoms (each forming four single bonds)
    6 hydrogen atoms (each forming a single bond)
    1 oxygen atom (forming two single bonds)

11. On paper draw a structural formula for each compound you have constructed, indicating how the nine atoms are connected. Compare your structures with those made by other classmates. After you are satisfied that all possible structures have been produced, answer these questions:

    a. How many distinct structures did you identify?
    b. What is the structural formula for each structure?
    c. Are all of these structures isomers? Explain.

12. Each compound you have identified possesses distinctly different physical and chemical properties.

    a. Recalling that "like dissolves like," which compound should be most soluble in water?
    b. Which compound should have the highest boiling point?

Part 3: Alkene-Based Polymers

In this part of the activity, you and your classmates will use models to simulate the formation of several addition polymers.

13. Build models of two ethene molecules.

14. Using information on page 217 as a guide, combine your two models into a dimer, or a two-monomer structure. What modifications in the monomer structure were necessary to accomplish this?

15. Combine your dimer with that of another lab team. Continue this process until your class has created a long-chain structure. Although your resulting molecular chain is not yet long enough to be regarded as a polymer, you have modeled the processes involved in creating a typical addition polymer.

16. Repeat Steps 13 and 14 for vinyl chloride and again for acrylonitrile. Then determine whether you can do the same for styrene. (See page 217 for the molecular structures for these substances.)

17. Assume the structures you built in Steps 15 and 16 became significantly larger. Give the name of each resulting polymer.

18. Are the chains you built linear or branched? How would this affect the properties of the polymer?

19. Make a polyethene chain that includes some cross-linking.

    a. Does this change the behavior of the model?
    b. How would the cross-linking change the properties of the polymer?

C.4 MORE BUILDER MOLECULES

So far, you have examined only a small part of the inventory of builder molecules that chemical architects have available. Now you will explore two important classes of compounds in which carbon atoms are joined in rings rather than in chain structures.

As a first step in doing this, picture a straight-chain hexane molecule, $CH_3-CH_2-CH_2-CH_2-CH_2-CH_3$. Next, remove one hydrogen atom from the carbon atom at each end. Then imagine those two carbon atoms bonding to each other. The result is the molecule known as cyclohexane.

$$CH_3CH_2CH_2CH_2CH_2CH_3$$

Hexane

Cyclohexane

Cyclohexane is a starting material for making nylon, an important and familiar petrochemical polymer. Cyclohexane is representative of the **cycloalkanes**, which are saturated hydrocarbons made up of carbon atoms joined in rings.

Another important class of hydrocarbon builder molecules is the **aromatic compounds**. Unlike cycloalkanes, aromatic rings are unsaturated. Aromatic compounds have chemical properties distinctly different from those of the cycloalkanes and their derivatives. The structural formula of benzene (C_6H_6), the simplest aromatic compound, is shown below. In the representation on the right, each "corner" of the six-carbon (hexagonal) ring represents a carbon atom with its hydrogen atom.

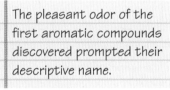

or

Although chemists who first investigated benzene proposed these structures, the chemical properties of the compound did not support their hypotheses. Recall that carbon-carbon double bonds ($C=C$) are usually very reactive. But benzene behaves as though it does not contain any such double bonds. A deeper understanding of chemical bonding was needed to explain benzene's puzzling structure.

Substantial experimental evidence indicates that all carbon-carbon bonds in benzene are identical. Thus its structure is not well represented by alternating single and double bonds. Instead, chemists often represent a benzene molecule in the following way:

> The pleasant odor of the first aromatic compounds discovered prompted their descriptive name.

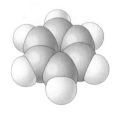

Benzene, C_6H_6

The inner circle represents the equal sharing of electrons among all six carbon atoms. The hexagonal ring represents the bonding of six carbon atoms to each other. Each corner in the hexagon is the location of one carbon atom and one hydrogen atom, thus accounting for benzene's formula, C_6H_6.

Although only small amounts of aromatic compounds are found in petroleum, large quantities are produced by petroleum fractionation and cracking. Benzene and other aromatic compounds are present in gasoline as octane enhancers; however, they are used primarily as builder molecules. Entire chemical industries (dye and drug manufacturing, in particular) are based on the unique chemistry of aromatic compounds.

C.5 BUILDER MOLECULES CONTAINING OXYGEN

In assembling molecular models with C, H, and O atoms, it is likely that you "discovered" one of the compounds depicted below:

$$CH_3-OH \qquad\qquad CH_3-CH_2-OH$$
Methanol (methyl alcohol)   Ethanol (ethyl alcohol)

As you can see, each molecule has an —OH group attached to a carbon atom. This general structure is characteristic of a class of compounds known as **alcohols**. The —OH is recognized by chemists as a **functional group**—an atom or group of atoms that imparts characteristic properties to organic compounds. If the letter R is used to represent all of the molecule other than the functional group, then the general formula of an alcohol can be written as

$$R-OH$$
Any alcohol

> The —OH alcohol functional group should not be confused with the hydroxide anion, OH⁻, found in ionic compounds such as sodium hydroxide, NaOH.

In this formula, the line indicates a covalent bond between the oxygen of the OH group and an adjacent carbon atom in the molecule. In methanol (CH_3OH), the letter R represents CH_3-; in ethanol (CH_3CH_2OH), R represents CH_3CH_2-.

The formulas and structures of two common alcohols are shown below. What does R represent in each compound?

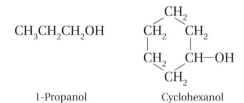

$$CH_3CH_2CH_2OH$$

1-Propanol   Cyclohexanol

Two other classes of oxygen-containing compounds, **carboxylic acids** and **esters**, are versatile and important builder molecules. The functional group in each class of compounds contains two oxygen atoms.

$$\begin{array}{cc} O & O \\ \parallel & \parallel \\ R-C-OH & R-C-OR \\ \text{Carboxylic acid} & \text{Ester} \end{array}$$

Note that both classes of compounds have one oxygen atom double-bonded to a carbon atom and a second oxygen atom single-bonded to the

(a)

Figure 20: *Familiar items that contain (a) alcohols and, on page 225, (b) carboxylic acids and (c) esters.*

same carbon atom. Ethanoic acid (a carboxylic acid that is better known as acetic acid) and methyl ethanoate (an ester more commonly called methyl acetate) are examples of these two classes of compounds.

(b)

$$H-\underset{\underset{H}{|}}{\overset{\overset{H}{|}}{C}}-\overset{\overset{O}{\|}}{C}-OH \qquad H-\underset{\underset{H}{|}}{\overset{\overset{H}{|}}{C}}-\overset{\overset{O}{\|}}{C}-OCH_3$$

Ethanoic acid   Methyl ethanoate

Some familiar things that include or involve the use of alcohols, carboxylic acids, and esters are shown in Figure 20. Builder molecules can be modified by adding other functional groups that include nitrogen, sulfur, or chlorine atoms. The rich variety of functional groups greatly expands the types of molecules that can ultimately be built by chemists.

(c)

C.6 CONDENSATION POLYMERS

Earlier you learned that many polymers can be formed from alkene-based monomers through addition reactions. Not all polymers are formed in this way, however. Natural polymers, such as proteins, starch, cellulose (in wood and paper), and synthetic polymers, including the familiar nylon and polyester, are also formed from monomers. But unlike addition polymers, these polymers are formed with the loss of simple molecules such as water when monomer units join. The term **condensation reaction** applies to this second type of polymer-making process, and the resulting product is called a **condensation polymer**.

One very common condensation polymer is polyethylene terephthalate (PET). This polymer is commonly used in large soft-drink containers. It also enjoys many other applications—as thin film for videotape and as the textile Dacron® for clothing, surgical tubing, and fiberfill. More than two million kilograms of PET are produced each year in the United States. Some examples of products made of PET are shown in Figure 21.

While some PET may be recycled, material that contains high quantities of additives, such as dyes or other polymers, is either landfilled or incinerated. The DuPont PetretecK process breaks down PET into its monomers, which can then be repolymerized into high-quality PET products. This technology decreases the need for new petroleum-based builder molecules as well as the quantity of PET being landfilled.

Condensation reactions can be used to make small molecules as well as polymers. In the next laboratory activity, you will use condensation reactions to produce esters. These reactions illustrate how organic compounds can be combined chemically to create new substances.

Figure 21 *Common uses of PET.*

C.7 CONDENSATION

Introduction

In this activity you will produce several petrochemicals through the reaction of an organic acid (an acid derived from a hydrocarbon) with an alcohol. The esters you will produce have familiar, pleasing fragrances. Many perfumes contain esters; the characteristic aromas of many herbs and fruits arise from esters contained in the plants.

> Organic acids are also called carboxylic acids. See page 224.

One example of the formation of an ester is the production of methyl acetate from ethanoic acid (acetic acid) and methanol in the presence of sulfuric acid:

$$\underset{\text{Ethanoic acid}}{CH_3-\overset{\overset{\displaystyle O}{\|}}{C}-OH} + \underset{\text{Methanol}}{H-O-CH_3} \xrightarrow{H_2SO_4} \underset{\text{Methyl acetate}}{CH_3-\overset{\overset{\displaystyle O}{\|}}{C}-O-CH_3} + \underset{\text{Water}}{H-OH}$$

To emphasize the roles of functional groups in the formation of an ester, a general equation can be written using the R notation you learned earlier.

> Recall that R stands for the "rest" of the molecule, or everything other than the functional group.

$$\underset{\text{Carboxylic acid}}{R-\overset{\overset{\displaystyle O}{\|}}{C}-OH} + \underset{\text{Alcohol}}{H-O-R} \xrightarrow{H_2SO_4} \underset{\text{Ester}}{R-\overset{\overset{\displaystyle O}{\|}}{C}-O-R} + \underset{\text{Water}}{H-OH}$$

Note how the functional groups of the acid and alcohol combine to form a water molecule, while the remaining atoms join to form an ester molecule. Sulfuric acid (H_2SO_4) acts as a catalyst—that is, it causes the chemical reaction to proceed faster without itself being used up.

Using a process similar to the reaction shown here, you will now produce the ester known as methyl salicylate.

Procedure

1. Prepare a water bath by adding about 50 mL tap water to a 100-mL beaker. Place the beaker on a hot plate, and heat the water until it is near boiling.

2. Obtain a small, clean test tube. Place 5 drops methanol into the tube. Next add 0.1 g salicylic acid. Then add 2 drops concentrated sulfuric acid to the tube. **CAUTION:** *Concentrated sulfuric acid will cause burns to skin or fabric. Add the acid slowly and very carefully.*

3. As you dispense these reagents, note their odors. **CAUTION:** *Do not directly sniff any reagents—some may irritate or burn nasal passages.* Record any odors you happen to note.

4. Place the test tube in the water bath you prepared in Step 1.

5. Using test-tube tongs, move the test tube slowly in the water bath in a small horizontal circle. Keep the tube in the water, and do not spill the contents. Note any color changes. Continue heating for three minutes.

6. If you have not noticed an odor after three minutes, remove the test tube from the water bath, hold the test tube away from you with the tongs, and wave your hand across the top of the test tube to waft any vapors toward your nose. Record observations regarding the odor of the product. Compare your observations with those of other class members.

7. Repeat the procedure using 20 drops pentyl alcohol, 20 drops acetic acid, and 2 drops sulfuric acid.

8. Repeat the procedure using 20 drops octyl alcohol, 20 drops acetic acid, and 2 drops sulfuric acid.

9. Dispose of your products as directed by your teacher.

10. Wash your hands thoroughly before leaving the laboratory.

Questions

1. Write the molecular formulas of the acid and alcohol from which you produced methyl salicylate. (*Hint:* You may need to look in a chemistry reference book.)

2. Write a chemical equation for the formation of methyl salicylate.

3. Repeat Questions 1 and 2 for the second ester you produced.

4. Describe the odors of the three esters produced in this laboratory activity.

5. Classify each of the following compounds as a carboxylic acid, an alcohol, or an ester:

   a. $CH_3-CH_2-CH_2-CH_2-OH$

   b. $CH_3-\overset{\overset{\displaystyle O}{\|}}{C}-O-CH_2-CH_3$

   c. $CH_3-\overset{\displaystyle \underset{\displaystyle CH_2-\overset{\overset{\displaystyle O}{\|}}{C}-OH}{|}}{CH}-CH_2-CH_3$

   d. $CH_3-\overset{\overset{\displaystyle O}{\|}}{C}-OH$

C.8 BUILDER MOLECULES IN TRANSPORTATION

Making Decisions

In this section, you have had a chance to survey some of the many and varied roles of petroleum products as building blocks for countless everyday objects. You can easily find examples of products built from petroleum in your surroundings, products that were unknown even a few decades ago. Plastics have many advantages over traditional materials, including favorable strength-to-weight ratios and recyclability.

In the TV advertisement that opened this unit, a newly designed automobile was described as "petroleum-free." Is that possible? Is it plausible? In this activity, you will consider this claim and evaluate how builder molecules are used in automobile construction.

1. List several automobile components that are commonly made of polymers and plastics.

2. Which of the polymer parts you listed are made from petroleum builder molecules? (*Hint:* It may be easier—and faster—to determine which parts are not petroleum-based.)

3. Try to think of a replacement material (not another polymer or petrochemical) for each component on your list.

Questions and Answers

4. Decide whether there are any disadvantages associated with each replacement material you have suggested. List the most important disadvantages for each component, or write "none" if the replacement material provides similar properties at a comparable cost. A table such as the one shown here will help you to organize your ideas and answers.

Automobile Component	Petroleum Based? (Yes or No)	Potential Replacement Material	Disadvantages of Replacement Material

Optimizing Technology

A polymer is a large molecule consisting of a chain of small molecules joined together in a repeating fashion. Chemists develop polymers for use as "ingredients" in products with unique physical and chemical properties. Polymer products can be lightweight, hard, strong, and flexible. They can also have special thermal, electrical, and optical characteristics. Products made of polymers come from a variety of industries, including textile and industrial fiber, plastics, communication, packaging, and transportation.

> **Chemical testing to produce a new product with new properties begins in a chemical laboratory.**

The big boom in polymer chemistry occurred during the early 1900s with the advent of polymer materials such as polyethylene, nylon, and synthetic rubber. Today, the main focus of polymer chemistry is improving existing technologies and developing new synthetic and processing technologies that are more efficient.

Polymer Chemistry: Fibers to Fit Your Needs!

Stacy Johnson is a Senior Research Chemist at the E.I. DuPont de Nemours and Company facility in Waynesboro, Virginia. Stacy's company makes LYCRA spandex, which is used in clothing such as underwear, hosiery, socks, swimwear, sportswear, and outerwear. LYCRA spandex is a strong elastomeric polymer that has the ability to stretch 400 to 800 times its original length without breaking. In the 1960s, LYCRA replaced much of the rubber commonly used in a variety of garments. Compared to rubber, LYCRA fibers have higher chemical stability and offer increased comfort and fit.

As a Senior Research Chemist, Stacy is responsible for developing new LYCRA products to meet the demands of the worldwide textile market. In the last decade, new chemical processes and additives have improved the properties of LYCRA, properties such as chlorine durability in swimwear and power enhancement for athletic performance in sportswear. In order to develop new products, Stacy must understand how market need and chemistry interact; that is, she must evaluate the various chemical routes to meeting that need and then implement the most effective one(s).

Chemical testing to produce a product with new properties begins in a chemical laboratory. An extensive "scale-up" process in which the same chemical reactions done in the laboratory are now reproduced in an industrial plant is needed to commercialize a new product. This process often requires several years of development. Thus an industrial chemist must not only know how to use chemistry to solve problems, but must also understand the larger picture that involves marketing and business.

Before earning a Ph.D. in chemistry at Pennsylvania State University, Stacy spent two years as a high school teacher in Athens, Texas. She feels that teaching was beneficial because through it she came to realize that chemistry, often an intimidating subject, can be made very exciting if it is explained in terms of its application to people's everyday lives. Stacy says, "Being a chemist is a very creative job. You are essentially building things with very tiny blocks that fit together in certain ways, and using the chemical rules and vocabulary you learned in school to describe what you built."

Preparing for a Career in Chemistry

1. Chemists are employed in a variety of fields. Conduct some research to find out what chemists working in the following areas do. (*Hint*: Go to the American Chemical Society Website at www.acs.org for help.)

   Food and Flavor Chemistry

   Forensic Chemistry

   Science Writing

   Chemical Education

   Analytical Chemistry

   Materials Science

   Medicinal Chemistry

   Biotechnology

   Hazardous Waste Management

   Environmental Chemistry

2. Science differs from technology. Review the definitions of the two words, and then identify several examples of each that are described in this Chemistry at Work feature.

3. The rolls in the picture above are spools of LYCRA yarn. Imagine stretching a spandex fiber with your hands and then releasing it. What types of interactions must the polymer molecules in the fiber have with each other in order to stretch many times the original length and then return to the original size and shape?

4. In addition to synthetic polymers, natural polymers are also used in clothing. List several fabrics made from natural polymers.

SECTION SUMMARY

Reviewing the Concepts

♦ The chemical combination of small, repeating molecular units called monomers forms large molecules called polymers. Polymers can be designed or altered to produce materials with desired flexibility, strength, and durability.

1. How many repeating units are found in each of the following structures?

   a. a monomer
   b. a dimer
   c. a trimer
   d. a polymer

2. List four examples of natural polymers and four examples of synthetic polymers.

3. What structural features make the properties of one polymer different from those of another?

4. List two techniques that can be used to modify the characteristics of a polymer. Explain each.

♦ Unsaturated hydrocarbons containing one or more double bonds are called alkenes. Unsaturated hydrocarbons containing a triple bond are called alkynes. Alkenes and alkynes are more chemically reactive than alkanes.

5. Why is the term unsaturated used to describe the structures of alkenes and alkynes?

6. Rank the following in order of decreasing reactivity: alkyne, alkane, and alkene. Explain your answer.

7. Draw an electron-dot diagram and structural formula for each of these molecules:

   a. propane   b. propene   c. propyne

♦ Addition reactions include the chemical combination of monomers to form a polymer, without any loss of smaller molecules.

8. Use structural formulas to illustrate how polymerization can result in the formation of polypropylene from propylene.

9. Explain why alkanes cannot be used to make polymers.

♦ Ring compounds include cycloalkanes and aromatic compounds.

10. In what ways are cycloalkanes different from aromatic compounds?

11. Draw a structural formula for both a saturated and an unsaturated form of C_4H_8.

12. How did aromatic compounds get their name?

13. Why is the circle-within-a-hexagon representation of a benzene molecule a better model than the hexagon with alternating double bonds?

♦ Functional groups—such as alcohols, carboxylic acids, and esters—impart characteristic properties to organic compounds.

14. a. Write the structural formula for a molecule containing at least three carbon atoms that represents (i) an alcohol, (ii) an organic acid, and (iii) an ester.

   b. Circle the functional group in each structural formula.
   c. Name each compound.

15. What does the R stand for in ROH?

- Condensation reactions involve the chemical combination of two larger molecules with the loss of a small molecule such as water.

16. Why is the word condensation used to describe the reaction that forms esters?

17. Draw the structural formula for the product of the following condensation reaction:

18. Acetic acid, CH_3COOH, and butyric acid, $CH_3(CH_2)_2COOH$, are common reactants in condensation reactions. Their structural formulas are shown at right. Predict the products of a condensation reaction between each acid and each of the following alcohols by providing the name and structural formula for the product:

a. methanol (CH_3OH)

b. ethanol (C_2H_5OH)

c. propanol (C_3H_7OH).

For example, the product of a condensation reaction between acetic acid and ethanol is ethyl acetate.

Connecting the Concepts

19. Chemical synthesis is one of many branches of chemistry. Test your skill at planning syntheses by identifying the missing molecule (represented by a question mark) in each of the following equations. (*Hint:* If you are uncertain about the answer, start by completing an atom inventory. Remember that the final equation must be balanced.)

a.

$$CH_3-CH_3 + ? \longrightarrow CH_3-CH_2Cl + HCl$$

b.

c.

20. Using a molecular model to represent each molecule, explain the differences between benzene and cyclohexane.

21. There is a saying that "Form follows function." How does this apply to the structure and properties of organic molecules?

22. List five products manufactured from cyclic compounds.

Extending the Concepts

23. Considering that petroleum is a natural resource, why aren't all petroleum-based polymers classified as natural?

24. Organic molecules that contain alcohol or carboxylic acid functional groups are often more soluble in water than similarly sized hydrocarbon molecules. Explain.

25. C_2H_6O has two isomers, one with a boiling point of 78 °C and the other with a boiling point of −24 °C.

a. Draw the structural formula for each isomer.

b. Which isomer has the higher boiling point? Explain your choice.

ENERGY ALTERNATIVES TO PETROLEUM

Because petroleum is a nonrenewable resource, its total available inventory on Earth is finite. Thus other sources of energy must be found in order to meet the needs of modern society. In this section, you will explore several alternatives to petroleum and also learn about other ways to power personal and commercial vehicles.

D.1 ENERGY: PAST, PRESENT, AND FUTURE

The Sun is our planet's primary energy source. Through photosynthesis, radiant energy from the Sun is stored as chemical energy. That is, green plants use the Sun's radiant energy to convert carbon dioxide and water into carbohydrates and oxygen gas. Thus green plants convert solar energy into chemical energy, which is stored in the bonds of carbohydrate molecules. When animals ingest and digest the plants, this chemical energy is released and used by the animals to form other organic molecules. Organic molecules found in plants and animals are called **biomolecules**.

Solar energy and energy stored in biomolecules are the key energy sources for life on Earth. Since the discovery of fire, human use of stored solar energy in wood, coal, and petroleum has been a major influence on civilization's development. In fact, the forms, availability, and cost of energy greatly influence how—and even where—people live.

In the past, abundant supplies of inexpensive energy were available. Until about 1850, wood, water, wind, and animal power satisfied the nation's slowly growing energy needs. Wood, then the predominant energy source, was readily available to most people, serving as an energy source for heating, cooking, and lighting. Water, wind, and animal power provided transportation and drove machinery and industrial processes. As demand for energy increased, primary fuel sources changed. Figure 22 illlustrates how U.S. energy sources have changed since 1850. In the next activity, you will explore how energy supplies and fuel use have shifted in the United States during the past 150 years.

> The total energy used in the United States in one week is approximately 2×10^{18} J.

> In 1850, the population of the United States was 23 million.

FUEL SOURCES OVER THE YEARS   Building Skills 8

As you can see from Figure 22, there has been a definite shift away from biomass (mainly wood, but also ethanol and waste) and toward fossil fuels—coal, petroleum, and natural gas. Use this figure to answer the questions that follow.

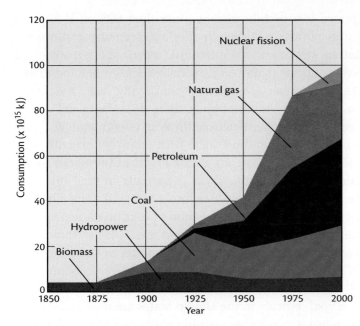

Figure 22: *Annual U.S. consumption of energy from various sources (1850–2000).*

1. a. Give the dates of the period during which biomass (mainly wood) supplied at least 50% of the nation's total energy.
    b. What were the chief modes of travel during that period?
    c. What factors might explain the decline in use of biomass after that period?
    d. What energy source increased in importance to replace biomass?

2. Compared with other energy sources, only a small quantity of petroleum was used as fuel before about the 1920s. What do you think petroleum's main uses might have been before that time?

3. Petroleum became increasingly important, and coal use began to decline, at about the same time.

    a. When did that occur?
    b. What could explain the growing use of petroleum after that date?

4. a. Which energy sources have become more important since 1975?
    b. What are the major uses of these energy sources?

5. a. Describe the trends in petroleum and coal use since 1975.
    b. What factors could explain these trends?

D.2 ALTERNATIVE ENERGY SOURCES

Everyday life in the United States requires considerable quantities of energy. As you just learned, the range of energy sources used in this country has indeed changed over time. As energy demands have accelerated, the nation has relied increasingly on nonrenewable fossil fuels—coal, petroleum, and natural gas. What is the future for fossil fuels, particularly petroleum?

The United States is a mobile society. More than 60% of U.S. petroleum is used for transportation. Although efforts to revitalize and improve public transportation systems merit attention, most experts predict our nation's citizens will continue to rely on personal vehicles well into the foreseeable future. And remember, even energy-conserving mass transit systems must have a fuel source. What options, then, does chemistry offer to extend, supplement, or even replace petroleum as an energy source?

Petroleum from tar sands and oil shale rock is an option with some promise. Major deposits of oil shale are located west of the Rocky Mountains. These rocks contain kerogen, which is partially formed oil. When the rocks are heated, kerogen decomposes into a material quite similar to crude oil. Unfortunately, vast quantities of sand or rock must be processed to recover this fuel. Moreover, current extraction methods use the equivalent of half a barrel of petroleum to produce one barrel of shale oil. Enormous amounts of water are also needed for processing—a problem where water is scarce.

> A metric ton of oil shale typically contains the equivalent of 80 to 330 L of oil.

Because known coal reserves in the United States are much larger than known reserves of petroleum, another possible alternative to petroleum is a liquid fuel produced from coal. The technology for converting coal to liquid fuel (and also to builder molecules) has been available for decades, having been used in Germany more than 50 years ago. Current coal-to-liquid-fuel technology is well developed in the United States. However, the present cost of mining and converting coal to liquid fuel is considerably greater than that of producing the same quantity of fuel from petroleum. But if petroleum prices increase, obtaining liquid fuel from coal—itself a nonrenewable resource—may become a more attractive option.

Petroleum replacement candidates are not limited to other fossil fuels. Certain plants, including some 2000 varieties of the genus *Euphorbia*, capture and store solar energy as hydrocarbon compounds rather than as carbohydrates. These compounds may prove to be extractable and usable as a petroleum substitute. Other alternative energy sources currently in use or under investigation include hydropower (water power), nuclear fission and fusion, solar energy, wind energy, biomass, and geothermal energy. Alternate approaches include constructing more energy-efficient buildings, vehicles, and machines, as well as using alternative fuels. All of these initiatives are intended to further reduce the need to burn petroleum.

You have learned about some alternatives to using petroleum as a fuel source. In the next section, you will examine some specific fuel alternatives.

D.3 ALTERNATIVE-FUEL VEHICLES

As you now know, personal vehicles consume a significant portion of the petroleum burned for fuel. In recognition of the limited nature of petroleum as a resource and in consideration of the emissions produced by petroleum-burning engines, alternative-fuel vehicles are being developed, tested, and used. What are some of these fuels, and how are they used to propel vehicles? What are the advantages and disadvantages of the various alternative fuels? The overview that follows will help you prepare your own automobile advertisement.

Compressed Natural Gas

Most passenger vehicles and buses can be converted to dual-fuel vehicles that run on either natural gas or gasoline. Natural gas, mainly methane (CH_4), is produced either from gas wells or during the processing of petroleum. Compressed and stored in high-pressure tanks, this product is commonly known as CNG (compressed natural gas). A refillable CNG tank, capable of powering an automobile up to 300 miles, can be comfortably installed in a car's trunk. Many CNG-powered vehicles are operating worldwide, particularly in government and mass transit fleets.

Among the advantages of CNG are wide availability and an 80% decrease (compared to gasoline) in carbon monoxide (CO) and nitrogen oxide (NO_x) emissions. However, refueling systems require a compressor, which increases the cost to $2000 to $4000 per vehicle.

Electric

Electric cars, including the fictitious ARL-600 vehicle featured in this unit, obtain their energy from a battery pack that is usually stored within the vehicle body and recharged with electricity at 120 or 240 volts. Powered by nickel-metal hydride (NiMH), lead-acid, or lithium-ion batteries, most electric vehicles can travel more than 100 miles on an eight-hour charge. The batteries provide energy to an electric motor, which turns the axle.

Figure 23 *An electric car being recharged.*

Although initially more expensive than gasoline-powered vehicles, electric vehicles do not require petroleum-based fuel, and their maintenance costs are lower. In addition, they do not produce any direct emissions. However, emissions may be produced some distance away at the electrical generating plant. And the limited lifetime of the batteries raises issues about recycling or disposal. The infrastructure for electric vehicles is largely in place in the form of the electrical power grid, but connections for recharging would need to be developed.

Fuel Cells

Fuel Cell

An emerging option for providing electricity to power vehicles is the fuel cell. Although fuel cells were in use before 1840, they did not become a practical energy source until the 1960s, when they were used in the U.S. Space Program. Any fuel containing hydrogen (such as methanol or natural gas) can be used in a fuel cell, but only hydrogen (H_2) can be used directly.

As shown in Figure 24, one common form of fuel cell converts hydrogen fuel and oxygen gas from the air into electrical energy and water, which is its only emission. From a chemical viewpoint, such an operating fuel cell represents another way to release and harness the energy involved in "burning" hydrogen gas:

$$2\,H_2 + O_2 \rightarrow 2\,H_2O + \text{Electrical energy.}$$

The fuel cell works by using porous electrodes that catalyze the reaction. In one common type of fuel cell, electrodes are in contact with a basic solution of potassium hydroxide, KOH(aq). At one electrode, hydrogen molecules (the fuel) react with hydroxide ions (OH^-), producing water and releasing electrons. The electrons flow from the fuel cell through an external circuit (where they do useful work), returning to the fuel cell at the other electrode. That second electrode catalyzes the reaction of oxygen gas (the oxidizer) with water and electrons to produce hydroxide ions, thus completing the electrical circuit. In the hydrogen-oxygen fuel cell, more water molecules are produced at one electrode than are consumed at the other.

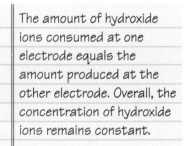

The amount of hydroxide ions consumed at one electrode equals the amount produced at the other electrode. Overall, the concentration of hydroxide ions remains constant.

Figure 24 *The chemical reaction 2 H₂ + O₂ → 2 H₂O + electrical energy takes place inside this fuel cell. See text.*

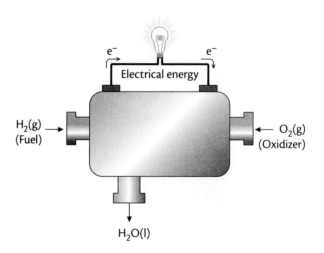

Fuel cells require no electrical recharging, eliminate or substantially reduce the release of air pollutants, and can obtain more useful power from a given quantity of fuel than can internal-combustion engines. However, challenges remain in developing fuel handling and processing options and reducing fuel-cell manufacturing and operating costs. The expense (in terms of high energy costs) required to obtain high-purity hydrogen (H_2) fuel for the type of fuel cell just described represents one of those key fuel-cell challenges.

Hybrid Gasoline-Electric

Some car designers believe that the hybrid gas-electric vehicle will best meet the needs of the consumer while reducing emissions and fuel costs. Hybrids come equipped with a small gasoline-burning engine as well as a battery-driven electric motor. The batteries are recharged while driving, partially through a conversion of braking friction into electricity. Hybrid vehicles typically achieve 70 miles per gallon and can travel more than 200 miles between fueling stops. Although hybrid vehicles produce the same kinds of emissions as fossil-fuel-burning vehicles, the quantity produced over a given distance is much smaller.

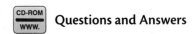 **Questions and Answers**

SECTION SUMMARY

Reviewing the Concepts

♦ **Continued reliance on petroleum as a fuel requires consideration of how to extend, supplement, or replace it as Earth's primary fuel source.** '

1. The supply of petroleum for both building and burning is limited. From the alternatives that have been discussed here, choose one for burning and one for building. Discuss the advantages and problems associated with the use of each substitute you have chosen.

2. Some energy authorities recommend exploring ways to use more renewable energy sources such as hydroelectric, solar, and wind power as replacements for nonrenewable fossil fuels.

   a. Why might this be a useful policy?
   b. Which of these renewable sources is/are least likely to replace fossil fuels at this time? Why?

3. Consider coal, oil shale, and biomass as possible fuel substitutes for petroleum. Which do you think might play the most useful role in for the future? Explain.

4. a. Of the two major uses of petroleum—as a fuel and as a raw material—which one is likely to be curtailed first if petroleum supplies dwindle?
   b. Give at least two reasons for your choice.

5. Although petroleum has been used for thousands of years, it is only in recent history that it has become a major energy source. List three technological factors that can explain this.

♦ **Alternative-fuel vehicles may be powered by compressed natural gas, electricity, or fuel cells.**

6. List an advantage and a disadvantage of each of the following alternative power sources for vehicles.

   a. compressed natural gas
   b. electric

   c. hydrogen fuel cell
   d. hybrid gasoline-electric

7. Alternative-fuel vehicles are commercially available today. What factors might discourage people from purchasing them?

Connecting the Concepts

8. U.S. reserves of oil shale are approximately 87 quads, the equivalent of 150×10^9 barrels of oil. Suppose that this represented our total source of oil, from which we maintained production of 8 million barrels of oil daily.

   a. How many years would this supply last?

   b. The population of the United States is about 2.8×10^8 persons, and the United States uses about 24 barrels (about 1000 gallons) of oil per person per year. At that rate of consumption, how long would the oil shale reserves last?
   c. Why is there a difference between the answers to Questions 8a and 8b?

9. We often consider alternative energy sources to be pollution free. Choose two alternative energy sources and evaluate them in terms of possible impact on the environment.

10. What are some ways in which individuals could reduce their total consumption of petroleum products?

11. A friend claims that because she does not own a car, she does not consume any fossil fuels. Evaluate this claim.

Extending the Concepts

12. Describe several possible changes that would occur in your community if wind and solar power devices were installed on a large scale to provide power.

13. How have the relatively low cost and easy availability of petroleum affected the search for and development of alternative energy sources?

14. What characteristics of chemical substances do chemists seek in good petroleum substitutes? Why?

15. Using Internet or library resources, identify some state or federal programs that encourage development of alternative energy sources.

16. World experts disagree about how long fossil-fuel supplies will last. Research and evaluate some of the current opinions.

PUTTING IT ALL TOGETHER
Getting Mobile

If you are an average high school student in the United States, you have already viewed about a half-million television commercials, many of them for automobiles. As alternative-fuel vehicles become available, advertisements similar to the one that begins this unit may become more common.

To make informed, intelligent consumer decisions, it is important to analyze the information conveyed in such advertisements. You practiced this skill earlier in the unit as you evaluated the "petroleum-free" claim in the ARL-600 TV ad. Now you will further develop the skill by creating and defending your own product claims. You will write and produce your own automobile commercial message.

DESIGNING, PRESENTING, AND EVALUATING VEHICLE ADS

Each team of students will create and present an advertisement featuring an imaginary but plausible vehicle that uses a specific type of fuel. The fuels you can choose are electric, gasoline, hybrid gasoline-electric, or hydrogen in a fuel cell. Your vehicle must use only the selected fuel. Your commercial message will be presented to the class for analysis and comment based on concepts introduced and discussed thus far in this chemistry class.

Each commercial message must meet certain specifications. Use these specifications to guide development of your advertisement as well as evaluate the presentations of your classmates.

- **Time** The presentation of the message for your vehicle should take no more than 60 seconds of "air time."
- **Scientific Claims** All scientific claims must be accurate. Because the type of fuel (energy source) is the unique characteristic of your vehicle, you should highlight it and explain it briefly. If appropriate, compare your vehicle to those that depend solely on petroleum for fuel. Include the nature and implications of emissions released by your vehicle, as well as a reference to its fuel efficiency.
- **Comfort/Design/Safety Features** Special features can enhance customer appeal as well as challenge your design creativity. Invent a name and model for your vehicle, and give it one or more special features that you think are particularly significant.
- **Presentation** In addition to presenting a commercial message with accurate claims, your goal is to stimulate interest and vehicle sales. To accomplish this, your commercial presentation should be organized, visually stimulating, and motivating. Your script should be concise and to the point, while still presenting all necessary details.

LOOKING BACK AND LOOKING AHEAD

This unit has illustrated once again how chemical knowledge can inform personal and community decisions related to resource use and replacement. Thus far in your *ChemCom* studies, you have focused on three distinctive types of resources: water, minerals, and petroleum. The next unit explores the importance of another resouce—the thin, virtually invisible envelope of gases that surrounds Earth's surface. As you will discover, the properties and behavior of the atmosphere can be easily overlooked, but issues and concerns arising from human interaction with it merit attention.

GLOSSARY

A

activity series
ranking of elements in order of chemical reactivity

addition polymer
polymer formed by repeated addition reactions at double or triple bonds within particular monomer units

addition reaction
reaction at the double or triple bond within an organic molecule

adsorbs
takes up or holds molecules or particles to the surface of a material

aeration
mixing of air into a liquid, as in water flowing over a dam

alcohols
nonaromatic organic compounds containing one or more –OH groups

alkali metal
highly reactive metal belonging to the first group of the Periodic Table

alkane
hydrocarbon containing only single covalent bonds

alkene
hydrocarbon containing a double covalent bond

alkyne
hydrocarbon containing a triple covalent bond

allotropes
two or more forms of an element in the same state that have distinctly different physical and/or chemical properties

alloy
solid solution consisting of atoms of different metals

anion
negatively charged ion

aqueous
solution in which water is the solvent

aquifer
structure of porous rock, sand, or gravel that holds water beneath the surface of Earth

aromatic compounds
ringlike compounds such as benzene that can be represented as having alternating double and single bonds between carbon atoms

atomic mass
mass of a particular atom of an element

atomic number
number of protons in an atom; distinguishes atoms of different elements

atomic weight
the average mass of an atom of an element as found in nature

atoms
smallest particles possessing the properties of an element

average value
See mean

B

bacterial action
conversion by bacteria of organic substances into simpler compounds

biomolecules
large organic molecules found in living systems

branched-chain alkane
alkane in which at least one carbon atom is bonded to three or four other carbon atoms

branched polymer
polymer formed by reactions that create numerous side-chains rather than linear chains

brass
alloy of copper and zinc; possesses properties different from those of either element

brittleness
a material's tendency to shatter under pressure

C

carbon chain
carbon atoms chemically linked to one another, forming a stringlike molecular sequence

carboxylic acid
organic compound containing the —COOH group

catalyst
substance that speeds up a reaction but is itself unchanged

cation
positively charged ion

ceramics
materials made by heating, or "firing," clay or components of certain rocks

chemical bond
force that holds atoms or ions together within a substance

chemical bonding
formation of chemical bonds

chemical change
change in matter resulting in formation of one or more new substances

chemical equation
symbolic expression that summarizes a chemical change, such as $2 H_2(g) + O_2(g) \rightarrow 2 H_2O(g)$

chemical formula
symbolic expression that represents the elements present in a substance together with subscripts indicating the relative numbers of atoms of each element

chemical property
a property that can only be observed or measured by changing the chemical identity of the sample of matter

chemical species
See species

chemical symbol
one- or two-letter abbreviation of the name of a chemical element, such as Fe for iron

chlorination
adding chlorine to a water supply to kill harmful organisms

coating
treatment in which a material is physically or chemically attached to a product's surface, typically for protection

coefficients
numbers in a chemical equation indicating the relative numbers of units of each species involved in the reaction

colloid
mixture containing particles that are small enough to remain suspended

combustion
burning

compound
substance composed of two or more elements in fixed proportions that cannot be broken down into simpler substances by physical means

condensation
conversion of a substance from the gaseous to liquid or solid state

condensation polymer
polymer formed by repeated condensation reactions of one or more monomers

condensation reaction
chemical combination of two organic compounds, accompanied by loss of water or other small molecule

conductor
material that allows electricity (or heat) to flow through it

confirming test
laboratory test that can confirm the presence of a particular chemical species

control
in an experiment, a trial that duplicates all conditions except for the variable being investigated

cracking
process in which hydrocarbon molecules from petroleum are converted to smaller molecules using heat and a catalyst

cycloalkane
saturated hydrocarbon containing carbon atoms joined in a ring

D

data
objective pieces of information, such as information gathered in an experiment

density
the mass per unit volume of a given material

diatomic molecule
molecule made up of two atoms

dissolves
involves a solute interacting with a solvent to form a solution

distillate
condensed products of distillation

distillation
method of separating liquid substances based on differences in their boiling points

doping
adding impurities to a semiconductor to modify its electrical conductivity

dot structure
See electron-dot formula

double covalent bond
bond in which four electrons are shared between two adjacent atoms

E

electrical conductivity
ability to conduct an electric current

electrometallurgy
use of electrical energy to process metals or their ores

electron
particle possessing negative electrical charge; found within atoms

electron-dot formula
formula of a substance or ion in which dots represent the outer electrons in each atom

electron-dot structure
See electron-dot formula

electronegativity
an expression of the tendency of an atom to attract shared electrons within a chemical bond

electroplating
deposition of a thin layer of metal on a surface by an electrical process involving oxidation-reduction

elements
fundamental chemical substances from which all other substances are made

endothermic
process requiring the addition of energy

energy level
expression of the relative potential energy possessed by electrons located within an atom

ester
organic compound containing the –COOR group, where R represents any stable arrangement of carbon and hydrogen atoms

evaporation
change of a substance from liquid to gaseous state

exothermic
process involving the release of energy

extrapolation
estimating a value beyond the known range

F

family (Periodic Table)
See group

filtrate
liquid collected after filtration

filtration
separation of solid particles from a liquid by passing the mixture through a material that retains the solid particles

flocculation
formation of an insoluble material suspended in or precipitated from a solution

fluoridation
addition of small quantities of fluoride to treated water supplies

formula unit
group of atoms or ions represented by a compound's chemical formula; simplest unit of an ionic compound

fraction (petroleum)
mixture of petroleum-based substances with similar boiling points and other properties; collected during distillation

fractional distillation
separating a mixture into its components by boiling and condensing the components sequentially

functional group
atom or group of atoms that imparts characteristic properties to an organic compound

G

gaseous state
state of matter having no fixed volume or shape

gasohol
a fuel produced by blending small quantities of ethanol with gasoline

green chemistry
chemical production methods that minimize or avoid the use of potentially harmful materials and the generation of potentially harmful pollutants

groundwater
water that collects underground

group (Periodic Table)
vertical column of elements in the Periodic Table; also called a *family*; members of a group share similar properties

H

hard water
water containing relatively high concentrations of calcium (Ca^{2+}), magnesium (Mg^{2+}), or iron(III) (Fe^{3+}) ions

heat of combustion
quantity of thermal energy released when a specific amount of a substance burns

heterogeneous mixture
mixture that is not uniform throughout

histogram
graph indicating the frequency or number of instances of particular values (or value ranges) within a set of related data

homogeneous mixture
mixture that is uniform throughout

hydrocarbon
molecular compound composed only of carbon and hydrogen

hydrometallurgy
metal processing methods involving treatment of rocks or ores with reactants in water solution

I

intermolecular forces
forces of attraction among molecules

ion
electrically charged atom or group of atoms

ion exchange
process of purifying water, which may involve the exchange of hard-water ions for other ions such as sodium (Na^+)

ionic compound
substance composed of positive and negative ions

isomers
compounds with the same molecular formula but different structural formulas

isotopes
atoms of the same element with different numbers of neutrons

K

kinetic energy
energy associated with motion

L

law of conservation of energy
energy can change form, but is not created or destroyed in a chemical reaction

law of conservation of matter
matter is not created or destroyed in a chemical reaction

liquid state
state of matter with fixed volume but no fixed shape

M

malleability
property related to a material's ability to be flattened without shattering

mass number
sum of the number of protons and neutrons in the nucleus of an atom of a particular isotope

mean

a number obtained by dividing the sum of a set of values by the total number of values in the set; also referred to as the *average value*

median

within an ascending or descending set of values, the number that represents the middle value, with an equal number of values above and below it

metal

a material possessing properties such as luster, ductility, conductivity, and malleability

metalloid

material with properties intermediate between those of metals and nonmetals

mineral

a naturally occurring solid substance; commonly removed from ores to obtain a particular element of interest

mixture

combination of materials in which each material retains its separate identity

molar heat of combustion

quantity of thermal energy released by burning one mole of a substance

molar mass

mass (usually in grams) of one mole of a substance or other chemical species

mole (mol)

an amount of substance or chemical species that is equal to 6.02×10^{23} units, where the unit may be atoms, molecules, formula units, electrons, or other specified entities; chemist's "counting" unit

molecule

smallest particle of a substance retaining the properties of the substance

monomer

compound whose molecules can react to form the repeating units of a polymer

N

natural water

untreated water gathered from natural sources such as rivers, ponds, or wells

neutron

particle possessing no electrical charge; found within the nuclei of most atoms

noble gas

very unreactive element belonging to the last (rightmost) group on the Periodic Table

nonconductor

material that does not allow electricity to flow through it

nonmetal

a material possessing properties such as brittleness, lack of luster, and nonconductivity (it acts as an insulator)

nonrenewable resource

a resource that will not be replenished by natural processes over the time frame of human experience

nucleus, atomic

dense, positively charged central region in an atom; composed of protons and neutrons

O

octane number

rating indicating the combustion quality of gasoline compared to the combustion quality of isooctane

OIL RIG

mnemonic device for remembering the definitions of oxidation and reduction: **O**xidation **I**s **L**oss, **R**eduction **I**s **G**ain (of electrons)

ore

rock or other solid material from which it is profitable to recover a mineral that contains a metal or other useful substance

organic chemistry

branch of chemistry dealing with hydrocarbons and their derivatives

oxidation

any process in which one or more electrons are lost by a chemical species

oxidation-reduction (redox) reaction

reaction in which oxidation and reduction occur

oxidizing agent

species that causes another atom, molecule, or ion to become oxidized

oxygenated fuel
fuel with oxygen-containing additives, such as methanol, that increase octane rating and reduce harmful emissions

P

parts per million, ppm
an expression of concentration, indicating the number of units of solute found within a million units of solution

percent composition
percent by mass of each component found in a material

Periodic Table
table in which elements are placed in order of increasing atomic number, such that elements with similar properties are placed in the same vertical column (group)

period (Periodic Table)
horizontal row of elements in the Periodic Table

petrochemical
any organic compound produced from petroleum or natural gas

physical change
change in matter in which the identity of the substance involved is not changed

physical property
a property that can be observed or measured without changing the identity of the sample of matter

polar
having regions of positive and negative charge resulting from uneven distribution of electrical charge

polyatomic ion
ion made up of two or more atoms

polymer
molecule composed very large numbers of identical repeating units

post-chlorination
addition of a low level of chlorine to treated water to prevent later bacterial infestation

potential energy
energy associated with position

pre-chlorination
addition of chlorine early in the water-treatment process to kill organisms

precipitate
insoluble solid substance that has separated from a solution

products
substances produced in a chemical reaction

proton
particle possessing positive electrical charge; found in the nucleus of all atoms

pyrometallurgy
use of thermal energy (heat) to process metals or their ores

R

range
difference between the highest and lowest values in a data set

reactants
starting substances in a chemical reaction

redox reaction
See oxidation-reduction reaction

reducing agent
species that causes another atom, molecule, or ion to become reduced

reduction
any process in which one or more electrons are gained by a chemical species

reference solution
solution of known composition used as a comparison in chemical tests

regeneration
process for renewing a material; renewing used ion-exchange resin by replacing hard-water ions in the resin with ions such as sodium (Na^+)

renewable resource
resource that can be replenished by natural processes over the time frame of human experience

S

sand filtration
separation of solid particles from a liquid by passing the mixture through sand

saturated hydrocarbon
hydrocarbon consisting of molecules in which each carbon atom is bonded to four other atoms

saturated solution
solution in which the solvent has dissolved as much solute as it can retain stably at a given temperature

semiconductor
solid substance, such as silicon, with electrical conductivity between that of conductors and nonconductors at normal temperatures

single covalent bond
bond in which two electrons are shared by the two bonded atoms

solid state
state of matter having a fixed volume and fixed shape

solubility
quantity of a substance that will dissolve in a given quantity of solvent to form a saturated solution

solute
dissolved substance in a solution, usually the component present in the smaller quantity

solution
homogeneous mixture of two or more substances

solvent
dissolving agent in a solution, usually the component present in the larger quantity

species
in chemistry, atoms, molecules, ions, radicals, or other well-defined entities

specific heat capacity
quantity of heat needed to raise the temperature of 1 g of a material by 1 °C

states
forms—gas, liquid, and solid—in which matter is found

straight-chain alkane
alkane consisting of molecules in which each carbon atom is linked to no more than two other carbon atoms

structural formula
chemical formula showing the arrangement of atoms and covalent bonds in a molecule, in which each electron pair in a covalent bond is represented by a line between the symbols of two atoms

subscript
character printed below the line of type used to indicate the number of atoms of a given element in a chemical formula; in H_2O, for example, the subscript 2 indicates the number of H atoms

substance
an element or a compound; a material with uniform, definite composition and distinct properties

supersaturated solution
solution containing a higher concentration of solute than a saturated solution at the given temperature

surface water
water on the surface of Earth

suspension
mixture containing large, dispersed particles that can settle out or be separated by filtration

T

tap water
fresh water drawn from plumbing lines

tetrahedron
a regular triangular pyramid; the four bonds of each carbon atom in an alkane point to the corners of a tetrahedron

turbidity
cloudiness

Tyndall effect
visible pattern caused by reflection of light from suspended particles in a colloid

U

unsaturated compound
organic compound with one or more double or triple bonds per molecule

unsaturated solution
solution containing a lower concentration of solute than a saturated solution at the given temperature

W

water cycle
repeating processes of rainfall (or other precipitation), run-off, evaporation, and condensation that continuously circulate water between Earth's crust and atmosphere

water softener
apparatus containing an ion-exchange resin; used to treat water by removing ions that cause water hardness

water treatment
purification processes applied to water before it is distributed for consumption